Attitudes Toward Local and National Government Expressed over Chinese Social Media

A Case Study of Food Safety

Douglas Yeung, Astrid Stuth Cevallos

For more information on this publication, visit www.rand.org/t/rr1308

Library of Congress Cataloging-in-Publication Data is available for this publication.
ISBN: 978-0-8330-9299-1

Published by the RAND Corporation, Santa Monica, Calif.

Cover: Image via imtmphoto/Fotolia

Preface

This report describes a psycholinguistic analysis of social media intended to explore how social media can provide a leading-edge perspective on how the Chinese public feels regarding domestic political issues (e.g., environment, food safety, local versus national government). The overarching purpose of the project was to understand the content and tone of messages transmitted via social media where those topics were being discussed. This initial exploration of Chinese social media also describes a methodology for analyzing Chinese social media, possibilities and challenges, required tools and steps, and comparisons with English-language social media. This research should be of interest to policymakers interested in public opinion regarding different levels of government and to researchers who wish to explore Chinese-language social media.

This study was made possible by a generous gift from the Cyrus Chung Ying Tang Foundation. Through this gift, the RAND Center for Asia Pacific Policy established the Tang Institute for U.S.-China Relations in 2007, within which this research was conducted.

RAND Center for Asia Pacific Policy

The RAND Center for Asia Pacific Policy (CAPP) is part of International Programs at the RAND Corporation. CAPP provides analysis on political, social, economic, and technological developments in and around the Asia Pacific. Through research and analysis, CAPP helps public and private decisionmakers solve problems, tackle challenges, and identify ways to make society safer, smarter, and more prosperous.

For more information on the RAND Center for Asia Pacific Policy, see www.rand.org/international_programs/capp or contact the director (contact information is provided on the web page).

Contents

APPENDIX

Figures

Tables

Summary

Social media are an increasingly common means of political expression. Analysis of them may help explain public opinion and perceptions of political events and movements. Psycholinguistic research—research on what language can reveal about psychological states, such as attitudes and emotions—has yielded insight on public opinion and mood during political events in Iran and Hong Kong. This work uses a computerized methodology to analyze social media text in China, specifically that regarding sentiment about local and national government and Western companies during the 2014 Husi scandal regarding expired meat in McDonald's and KFC restaurants.

In China, social media have expanded public debate on issues relating to governance. Some believe their use has helped effect meaningful policy changes, while others contend it has merely provided a safety valve for citizens to air their grievances without threatening the regime. The number of Chinese Internet users is enormous—nearly 650 million, or 48 percent of the population, a higher number than in India, and a higher percentage than in India, Thailand, and Vietnam. Among these users, microblogging is widespread, with Sina Weibo being the most active microblogging platform.

The number of Chinese microblog users reached a peak of 309 million in 2012; the number has decreased somewhat since then as the government has increasingly monitored content. Social media platforms popular elsewhere in the world, such as Facebook and Twitter, are not widely used in China, given government blocking of these sites. Still, many Chinese netizens do access Twitter through a proxy server or other means to "climb over the Great Firewall." Estimates of the number of Chinese citizens using Twitter vary widely, from 10,000 to 35 million.

Food-safety issues offer insight into how social media might help gauge public opinion in China. Food safety became a widespread concern in 2008, when a state-owned company was discovered to have contaminated milk formula with melamine. In subsequent years, the proportion of Chinese considering food safety to be a substantial problem increased sharply—particularly among high-income earners and younger Chinese most likely to be active on Sina Weibo. Analysis of social media postings on these issues, and particularly on the Husi scandal as it unfolded in July and August of 2014, can help identify dynamics of Chinese opinion on them.

Methods

We selected samples of Chinese-language microblog posts that discussed food safety in general, and the Husi incident in particular. We collected 2,184 Chinese-language tweets from June 1 to August 31, 2014, and 537 weibos from August 26 to September 8, 2014. Unfortunately, we were able to collect weibos only from the start of data collection, and not retroactively, as we did with tweets. Nevertheless, the samples may offer insights on how differing audiences—a more elite one that will circumvent barriers to use Twitter and a mass one possibly reflecting broader Chinese opinion—view these issues and responsibility for them.

In addition to analyzing postings for references to local and national governments and to Western companies, we used the automated software to identify pronouns, emotion words, and responsibility words. Use of first-person singular pronouns (e.g., I, me, my) suggests self-focus, which is associated with negative states, such as threat, depression, or insecurity. First-person plural pronouns (e.g., we, us, our) suggest the author has a sense of shared group or community. Third-person pronouns (e.g., he, she, they) suggest the author does not identify with the group he or she is discussing. Emotion words can indicate whether the author feels positively or negatively about a topic. Using responsibility words when discussing a particular entity may suggest that an author associates responsibility for food safety with that entity.

Results

Twitter Analysis

Changes in post-Husi sentiment about food safety suggest that Chinese opinion is attuned to both local and national government responses. During the week the Husi incident occurred, negative sentiment in tweets discussing food safety increased sharply, while positive sentiment decreased. By the time the national government asked foreign fast-food chains to post supplier information on their Chinese websites, negative sentiment on food safety had returned to pre-incident levels. Negative sentiment on the Husi incident in particular, however, remained elevated for several weeks.

Posts on the incident also had high rates of use for first-person singular pronouns, further suggesting negativity regarding the scandal immediately after its occurrence. When Shanghai police reported the arrest of executives held responsible for the tainted meat, use of other pronouns in Twitter posts peaked, suggesting a distancing from these executives.

Twitter frustration regarding the incident may have been directed less toward the local government and more toward the national government and U.S. companies. In particular, Twitter users were more likely to use third-person pronouns in discussing the national government and U.S. companies. Netizens also used twice as many negative emotion words regarding the national government and U.S. companies as they did regarding the local government. Analysis of responsibility words indicates that Twitter users viewed local governments as responsible for, or perhaps particularly effective in, punishing illegal behavior. By contrast, the term for corruption appeared almost twice as often in tweets about the national government as in tweets about the local government. This may reflect popular attitudes about corruption at national or local levels or that people spoke frequently about corruption in China or about the whole country in general.

Weibo Analysis

Weibo users appear to have felt a greater sense of community and less depression or insecurity regarding the national government. In discussing food safety, they appear to regard the national government more positively and local government more negatively than Twitter users.

First-person singular pronouns were used in weibos discussing local government nearly twice as often as in those discussing the national government. By contrast, weibos about the national government used first-person plural pronouns about five times more often than did those about the local government—suggesting that users felt a greater sense of kinship in discussing food safety in China at large than at city or provincial levels.

At the same time, Weibo users directed more expressions of anger toward the national government than local governments. Weibos referencing the nation or the national government on food safety contained more than four times the proportion of words conveying anger as those referencing municipal or provincial authorities on these issues. They also contained three times the proportion of words conveying sadness as weibos referencing local government.

While very few weibos in the sample referenced America or U.S. companies—likely because of the delay in data collection after the Husi incident—those that did contained far more posts about responsibility or blame than did those referring to Chinese entities. Further manual content analysis may be needed to disentangle specific interpretations of these and other results.

Conclusion

The divergences between Twitter and Weibo users may reflect the attitudes of different communities. People tweeting in Chinese may be more politically active and critical of the national government, given the additional effort required to access Twitter. Weibo users could be more representative of overall political views.

Alternatively, the greater positive sentiment toward the national government on Weibo may reflect the national government's control over Chinese media. The correspondence between public opinion as expressed on Weibo and the national government's interest in this case may reflect a public-relations campaign by the national government to deflect governance scandals onto local officials and Western companies. At the same time, posters used more emotional terms about the national government than about other entities—suggesting the national government was unable to deflect this scandal completely, or that users view food safety as a broad issue for the national government to regulate.

Future studies can extend investigation of Chinese domestic platforms and social media platforms and address challenges of automated analysis. Broadening key words (e.g., by including *meat quality* or *fast food* in analysis of food quality) could capture a wider range of discussion. Future studies might also explore more-rigorous methods for selecting words used to categorize tweets. They might also compare Facebook and Twitter users with users of more-common Chinese platforms.

Other domestic issues, such as pollution and environmental concerns, may warrant similar investigation. Netizens may hold divergent views about national and local responsibility for these issues. They may, for example, applaud the national government for issuing mandates to reduce pollution but blame local governments for ignoring new regulations—a common practice, given the lack of incentives to police state-owned companies.

Social media research of Chinese political issues may yield insights used in unintended ways—including furthering of social control. Researchers should consider the implications of how policymakers may use their findings.

Acknowledgments

Mike Lostumbo and Seth Jones provided support and valuable guidance to conduct this research. Zev Winkelman, Christopher Skeels, and David Manheim assisted with data collection and management. The Academia Sinica Lexicon/Corpus Group and the Chinese LIWC Dictionary Group provided assistance with C-LIWC. Finally, Rafiq Dossani, Rogier Creemers, and an anonymous reviewer provided thoughtful comments and suggestions that helped greatly improve this report.

Abbreviations

CCP	Chinese Communist Party
C-LIWC	Chinese Linguistic Inquiry and Word Count
LIWC	Linguistic Inquiry and Word Count
PX	paraxylene
SC-LIWC	simplified Chinese Linguistic Inquiry and Word Count

CHAPTER ONE

Social Media Use in China for Political Expression

The use of social media has become increasingly common in political events across the world, such as protest movements or mobilizations. Accordingly, analysis of such social media use may help explain public opinion and perceptions that may have resulted in these movements, or that provide insight into them. Previous psycholinguistic research—that is, examining what language can reveal about psychological states such as attitudes and emotions—investigated how social media can help in understanding public opinion and mood during ongoing political events in Iran (Elson et al., 2012) and Hong Kong (Yeung and Cevallos, 2014). Building on this work, which used a computerized methodology developed to analyze social media text, we explored attitudes and discussion in Chinese social media. We investigated sentiment about local government, national government, and Western companies as expressed by Chinese-language users on the microblogging platforms Twitter and Sina Weibo. We chose food safety as the specific topic to investigate and, in particular, used the July 2014 Husi scandal around expired meat in McDonalds and KFC (described in further detail below) as a case study.

Food safety touches on several important issues in Chinese domestic politics, including the role of the press (and the public) as a watchdog and the challenges of government regulation in a system where responsibility for enforcement is diffuse, the rule of law is weak, and local officials lack incentives to inspect and penalize profitable local businesses (Olesen, 2011). Over the past decade, food-safety issues involving contaminated baby formula and pork have alarmed Chinese consumers, fueling fear and distrust of the domestic food supply (Jiang, 2008; Barclay, 2011). Some news articles suggest that, in the wake of these scandals, Chinese officials are permitting a broader range of media coverage and public discussion about food-safety incidents in an attempt to better monitor China's burgeoning food industry (Olesen, 2011). As a result, food-safety may also be a good way to investigate Chinese social media. For example, highly publicized food-safety scandals in recent years—such as the Husi incident, which exposed the use of expired meat in fast-food companies—provide specific, measurable events that should elicit social media conversation.

The research described in this report sought to explore whether and how Chinese social media users discussed the Husi incident and food safety, and whom (e.g., local Chinese government, national Chinese government, Western companies such as those involved in the food-safety scandals) they believed should bear responsibility for the Husi incident in particular, and for food safety more generally. To address these questions, we conducted an exploratory study of Chinese-language social media, collecting online posts from both Sina Weibo and Twitter. Given limited resources, we conducted qualitative rather than more-rigorous statistical or content analyses. The overarching aim of this research was to demonstrate the capabilities of our computerized methodology for social media analysis (e.g., Elson et al., 2012) to uncover insights regarding attitudes toward local and national government in Chinese domestic politics. This report also explains some of the challenges and limitations of conducting research on Chinese social media texts. The study findings and method of assessing Chinese-language social media may be of interest to Chinese-language analysts or to policymakers seeking a greater understanding of Chinese public opinion regarding domestic politics.

Social Media Use in Contemporary China

In China, social media have expanded public debate on issues relating to governance. Certainly, there is disagreement about the nature and role of social media in the Chinese political space; while some believe that social media use has helped effect meaningful policy change in China, others contend that it has merely provided a safety valve for citizens to air their grievances without actively threatening the Communist regime (Xiao, 2011). In recent years, changing Chinese policies toward social media have also altered the possibilities for civic action and association (Creemers, 2015). Nevertheless, social media appear to have become an important tool for expression and popular mobilization, especially with regard to exposing and censuring local government transgressions.

Who Is Using Social Media in China?

As of December 2014, nearly 650 million Chinese citizens (almost 48 percent of the Chinese population) used the Internet (China Internet Network Information Center, 2015, p. 25). For the sake of comparison: India, whose population is roughly the same size as China's, has 243 million Internet users (just 19 percent of the population), and Thailand, whose per capita GDP is roughly similar to China's, has 19 million Internet users (29 percent of the population). China's Internet penetration rate is closer to that of fellow authoritarian (and communist) regime Vietnam, which has 40 million Internet users (43 percent of the population) (World Bank, undated; Internet Live Stats, 2014). Most Chinese netizens (as many Chinese Internet users call themselves) are young (56 percent are under the age of 30), male (56 percent), educated (52 percent have at least the equivalent of a high school diploma), urbanites (slightly more than 70 percent live in cities), and have above-average incomes (53 percent have annual incomes of 24,000 RMB—around $3,900—or more, while the average annual household income in China is just 13,000 RMB, or around $2,100). A little more than 85 percent of Chinese netizens use their mobile phones to access the Internet (China Internet Network Information Center, 2015, pp. 28–35).[1]

[1] The China Internet Network Information Center (CNNIC) is China's domain-name registrar and a research center that currently falls under the purview of the Cyberspace Administration of China.

In particular, microblogging (or *weibo*) in China is widespread, and there are many indigenous platforms (e.g., Sina Weibo, Tencent Weibo, Sohu Weibo, Netease Weibo, Fanfou, and Digu), though Sina Weibo (also referred to simply as Weibo), launched in August 2009, is the most active of these platforms. Weibo platforms are similar to Twitter in that they allow users to publish short messages of 140 characters to a public audience. But because most Chinese words consist of just one or two characters, Chinese microblog posts can convey much more information than comparable English tweets.

Microblog users climbed to a peak of 309 million in 2012, meaning that 55 percent of Chinese Internet users had a weibo account (China Internet Network Information Center, 2014, p. 48). Since then (for reasons explored below), the number of weibo users has declined nearly 20 percent to 249 million at the end of 2014 (representing just 38 percent of Chinese Internet users) (China Internet Network Information Center, 2015, p. 56). Most of these users are young (53 percent are under the age of 24, and 90 percent are under the age of 34) and highly educated (71 percent have at least a college degree) (Bai, 2014). While Sina Weibo's official statistics peg the number of female and male Weibo users as equal, other researchers suggest that the percentage of male users is slightly higher—between 52 percent and 57 percent of all Weibo users (Ng, 2013). Most users were also well off: 48 percent made more than 36,000 RMB, or around $5,900, annually (Millward, 2012). As might be expected, the majority of Weibo users are urbanites—Beijing municipality (total population of an estimated 22 million) and Jiangsu province (home to an estimated 80 million Chinese citizens, including around 24 million Shanghai residents) each boast more than 15 million users, and Guangdong province (home to 105 million Chinese citizens) has more than 30 million users (Sina Weibo Data Center, 2012)—in fact, more than 20 percent of Weibo posts come from Guangdong alone (Ng, 2013).

In China, social media platforms popular in the United States and around the world, such as Facebook and Twitter, are not as widely used as one might expect. Concerned about the potential for rapid, uncensored communications via social media platforms to provoke organized protest, the Chinese government began blocking access to

Twitter, Facebook, and other social media sites in June 2009, around the 20th anniversary of the Tiananmen Square protests. Still, Chinese netizens may—and some do—access Twitter. But in order to do so, they must use a proxy server or VPN to "climb over the Great Firewall" (*fan qiang*) (see Tkacheva et al., 2013, pp. 93–117; and Robinson, Yu, and An, 2013). As a result, it is impossible to determine exactly how many Chinese citizens use Twitter.[2] Estimates vary widely: one controversial study pegged the number of Chinese citizens actively using Twitter at 35 million, which would be Twitter's largest global market (Ong, 2012), but at least two later analyses—and the approximation of one well-known Chinese Twitter user, Internet freedom activist Michael Anti—have suggested that the number of Chinese users is actually somewhere between 10,000 and 100,000 (Russell, 2012; Zhai, 2013; Ng, 2014). Because of the difficulty in obtaining reliable statistics about Twitter usage in China, the demographics of China's Twitter users are unknown. Given the additional effort and steps required for Internet users in China to access Twitter, while China's Twitter users may represent a small fraction of Chinese Internet users, they may also be more likely to access Twitter to communicate views or participate in discussions that might otherwise be censored on Chinese social media platforms. Indeed, much of the Chinese dissident community—including controversial artist and provocateur Ai Weiwei and Internet freedom activist and journalist Michael Anti (Zhao Jing)—is active on Twitter (Goodman, 2012). A 2011 report on the use of Internet filter circumvention tools (such as VPNs or proxy servers) among international bloggers similarly suggests that a small percentage of the general Internet population uses such tools, but that those who do are more likely to be highly connected and influential individuals capable of spreading information and opinions to a broader audience (Roberts et al., 2011). While this barrier to entry may result in selection bias and thus skewed demographics, it may be likely that demographic

---

[2] Censorship and Internet traffic control may not be the only reasons few Chinese netizens use Twitter or Facebook; some researchers suggest that the Chinese equivalents—like Tencent Weibo, Sina Weibo, and Renren—are simply better at serving the needs of Chinese Internet consumers. For more, see Sullivan, 2012.

trends for Sina Weibo (e.g., urban, educated) also apply in the case of Twitter users in China.

Censorship, Anonymity, and the Shifting Landscape of Social Media in China

As weibo platforms have become increasingly popular, they also have played a more significant role in political expression and mobilization in China. Given the potential for widely used communication tools such as Sina Weibo to help mobilize collective action against the government, Beijing has sought to control social media to reduce the likelihood that activity on Sina Weibo will generate an organized, regime-threatening political movement. Susan Shirk, an expert on Chinese politics, suggests that China's leaders recognize that individuals who take to the Internet to express their frustrations are "more likely to take the greater risk of participating in, or organizing, mass protests" (Shirk, 2011). As a result, at the regime's behest, company-employed content censors quickly remove sensitive content. In general, posts about "real-world events with collective action potential" are far more likely to be censored than expressions of discontent, suggesting that authorities seek to prevent would-be protestors from organizing over social media but are willing to allow netizens to voice their dissatisfaction (King, Pan, and Roberts, 2014, p. 1; see also King, Pan, and Roberts, 2013). In 2012, in an effort to gain control over potential regime-threatening discussion on Sina Weibo, government officials enacted two new regulations: one required netizens to register with their real names when opening accounts, and the other required Sina Weibo to monitor the posts of high-profile bloggers with more than 100,000 followers and eliminate any questionable posts within a five-minute window after posting (Tkacheva et al., 2013). As some newspaper articles note, measures to limit the power of influential bloggers appear to have become even more aggressive since Xi Jinping officially took power in March 2013 (Chin and Mozur, 2013).

While Sina Weibo is still widely used, its previously undeniable status as the social media platform of choice for expression and mobilization on political issues may be less certain. Nearly 10 percent of Sina Weibo users left the platform in 2013. Even among the remaining users, activity has dropped—23 percent of users have reduced their usage of Weibo, but only 13 percent have increased their usage (China Internet Network Information Center, 2014, p. 66). Similarly, a recent *Telegraph* study suggests that even Weibo's most-active users made 74 percent fewer posts in December 2013 than in March 2012 (Moore, 2014). That users are posting less frequently on Weibo or leaving the service entirely may be in response to two trends: first, an August 2013 government crackdown on high-profile Weibo users, and second, the rise of alternative social media platforms—some of which may be more difficult to censor—such as the mobile instant-messaging app WeChat (*weixin*). Nevertheless, Sina Weibo remains widely used, serving as an important channel for public engagement in governance. Weibo reported in June 2014 that it had a total of 157 million monthly active users and 70 million daily active users (Sina Weibo, 2014).[3] Perhaps to capitalize on the size and breadth of this audience, some local Chinese governments have created their own Weibo accounts to communicate more effectively with their constituencies (Schlaeger and Jiang, 2014).

The Use of Social Media as a Tool for Political Expression and Mobilization

In addition to its other uses, Weibo provides a platform for netizens to spread information about unfolding crises (especially those that involve official attempts to conceal corruption or mistakes), mobilize popular support for government responses to such events, and express criticisms of government policies. In a few cases, despite the weak nature of civil society in China (see Jiang, 2010, p. 117),[4] netizen engagement within the online Weibo community has led not only to action offline, but also to policy outcomes. This is particularly true in instances where microbloggers have brought a local issue to national attention, compelling authorities to respond, as in these two well-known cases:

---

[3] Note that there is no indication of whether or not these users are located in mainland China.

[4] Jiang argues that "China has not yet developed a true civil society in the sense of being based on associations wholly independent of the state" (2010, p. 117).

- In July 2011, a Sina Weibo user broke news about a train collision in Wenzhou that claimed the lives of more than 40 people. The subsequent flurry of activity on Weibo exposed efforts to bury the wrecked train cars in an attempt to impede a thorough investigation into the causes of the crash. Netizens uncovered systemic corruption and criticized the government for prioritizing rapid modernization over safety, reframing the tragedy in such a way that the regime was forced to act by dismissing the Railway Minister and convicting him of corruption (Tkacheva et al., 2013, pp. 104–107).
- In the summer of 2011, residents of Dalian took to Sina Weibo to organize a protest to close a paraxylene (PX) plant located near the city center. Concerned about the adverse health effects caused by overexposure to PX, a chemical used to manufacture plastics, more than 12,000 protesters flooded the city's streets after tropical storms caused damage to protective barriers around the chemical plant. Although censors tried to stop the spread of information and calls for mobilization on Weibo, organizers used the medium swiftly and effectively. On the same day that the protest occurred, officials closed the factory in question (Tkacheva et al., 2013, pp. 107–110).

These examples show how savvy netizens—usually from the young and well-educated urban middle class—have used Sina Weibo to catapult local abuses of power and other political grievances onto the national stage, where authorities have greater difficulty containing the spread of dissent and thus may be more likely to respond to such public pressure. Given China's lack of multiparty elections and other democratic mechanisms of government accountability, there is some evidence the central government believes social media can play an important role in monitoring corruption. In late 2014, the Central Discipline Inspection Committee attempted to capitalize on social media as a mechanism for information collection—while minimizing the platform's risk of collective action—by creating its own controlled, online platform for soliciting reports of corruption (Creemers, 2015). That interactions on social media might serve as a tool by which the central government can observe and control local governments underscores the consequences of this research into netizens' use of Sina Weibo to express their grievances with regard to food safety crises caused by systemic regulatory lapses.

Food Safety Regulation in China and the Husi Incident

Government Regulation and Local Versus National Government Responsibilities

At the authorities' behest, microblogging platforms censor individual posts that could lead to organized protests. Yet microblog posts expressing dissatisfaction are generally tolerated. While some find support for the "social media as safety valve" theory in this dichotomy (Xiao, 2011), there is another possible explanation: In China's highly decentralized political system, public expressions of grievances via social media may help the national government monitor and police local officials.[5]

Local officials have a historical tendency to disregard the central authorities in distant Beijing, as alluded to in the traditional Chinese saying, "the mountains are high, and the emperor is far away." Indeed, many Chinese believe that social unrest in China occurs when corrupt or incompetent local officials fail to implement well-intentioned central government directives. Chinese leaders have exploited this perception to deflect complaints onto local officials, adopting a strategy that Cheng Li of the Brookings Institution calls "think national, blame local" (Li, 2006; see also Nathan, 2003).[6] In fact, some analysts argue that the authoritarian regimes such as China's countenance and even encourage public protest—both virtual and physical—in order

[5] For more on the central government's difficulties in reining in local officials and the problems such central-local tensions pose to the regime, see Fewsmith, 2013.

[6] The Chinese public's perception, or assumption, of the central government's benevolence is not new, dating back to the days of Imperial China. Yet Chinese leaders continue to cultivate this perception, as it instills a measure of allegiance to the regime in the abstract; this allegiance may help to ensure that political complaints remain local and do not develop into comprehensive calls for regime change. For more on the history of this perception, see O'Brien and Li, 2006, especially pp. 42–47.

to monitor discontent, keep local officials in check, and enact policy changes where necessary (Johnson, 2004 and 2011; Lorentzen, 2013). In this way, then, expressions of public discontent and even protest can serve to strengthen rather than weaken the central government.

China's legislature, the National People's Congress, is in charge of the Food Safety Law—which, as of this writing, is in a revision process that incorporates public consultation—that defines national food-safety standards and regulations (Balzano, 2014 and 2015). But it is municipal- and provincial-level food and drug administration officials who must inspect and hold local businesses accountable for meeting those standards. When a food-safety scandal breaks, whom do Chinese netizens hold responsible for regulatory lapses? The answer to this question could provide further insight into the discussion about the relationship between the national and local governments in authoritarian states such as China.

Food Safety as an Issue of Public Concern

Food safety became an issue of widespread public concern in China in the wake of the Beijing Olympics in 2008, when state-owned Sanlu Group, the country's largest domestic milk-powder company, was discovered to have contaminated milk formula with melamine, a toxic and illegal chemical additive that artificially enhances protein content readings during testing. The adulterated infant formula sickened 294,000 infants and caused six infant deaths. Two men received the death sentence for their involvement in the scandal—one for manufacturing and selling melamine, and one for adding melamine to the milk powder. In addition, the president and CEO of Sanlu was sentenced to life in prison and fined 20 million RMB (nearly $3 million) (Barboza, 2009).

Since then, the Chinese public's attention to food safety as a public-health and safety issue has grown considerably. According to Pew Research Center polls conducted in 2013, 38 percent of Chinese consider food safety a "very big problem," up from 12 percent in 2008. The overall percentage of Chinese who consider food safety a "very big problem" or "moderately big problem" increased 26 percentage points from 49 percent in 2008 to 76 percent in 2013 (Pew Research Center, 2013). Urban Chinese, high-income earners, and young Chinese under the age of 30 were most likely to identify food safety as a serious concern. Notably, these are the same demographic groups that are most active on Sina Weibo.

Although Weibo did not exist yet at the time of the 2008 Sanlu melamine crisis, it served as a platform for information dissemination and expression of outrage in two food safety scandals that emerged in Shanghai in 2013. When, in March, nearly 6,000 diseased pig carcasses were found floating in Shanghai waters, dumped by farmers into the Huangpu River upstream of the town of Jiaxing, netizens took to Weibo, often using dark humor and sarcasm to express their frustration. For example, one netizen wrote, *"We are the generation that grew up drinking melamine milk, eating broiler chicken and sewage oil. Our immune systems have been well trained. Shanghainese people are not easily defeated by some dead pigs"* (Lu, 2013). A few months later, in May, officials arrested 63 people for selling $1.6 million of fox, mink, and rat meat chemically treated to taste and appear like lamb meat in Shanghai and Jiangsu province over the course of several years (Kaiman, 2013). Netizens were similarly outraged, often posting vomiting emoticons in response to the news. One netizen wrote, *"Just imagine—you think you're eating some lamb skewers, but what you are actually eating is some furry rats. How will I dare eat lamb again?"* (Funakoshi, 2013). It was in this context that online discussion about the Husi incident took place.

Time Line of the Husi Incident

On July 20, 2014, an undercover investigative report by Dragon Television uncovered Chinese workers at Shanghai Husi Food Company, a branch of American-owned meat producer and supplier OSI Group, using expired meat in McDonald's and KFC products and tossing meat that had fallen on the floor back into processors. Shanghai regulators subsequently discovered that Shanghai Husi forged the production dates on over 4,000 batches of beef patties; the meat, which had a nine-month shelf life, was stamped with a January 2014 production date but had actually been produced in May 2013 (Reuters, 2014). On July 21, the Shanghai Husi scandal was the "most viewed topic thread on Weibo, with over 27 million people deploring foreign fast food, China's food safety monitoring, and local media's tendency to make news by going after well-known consumer companies" (Kuo, 2014). On Sina Weibo, the scandal was referred to, alternately, as "Husi incident" (*fuxi shijian*), "Expiration Gate" (*guoqi men*), and even "Stinky Meat Gate" (*chourou men*).[7]

As citizens expressed their frustration and disgust on social media, local and national government regulators scrambled to respond, and the American businesses implicated in the scandal—including McDonald's and KFC—moved to sever ties with Shanghai Husi and stopped offering some meat products at their franchises. Table 1.1 describes the time line of the government's responses to the crisis.

Our case study differs from many of the previously discussed examples in which Chinese netizens exposed improper conduct, corruption, and cover-ups. Here, netizens neither broke the story of regulation failure at Shanghai Husi nor organized real-life protests to effect change. Thus, it might appear that the role of social media in this case is tangential because the traditional media covered the scandal first. However, netizens' *responses* to the scandal on social media are the object of our study, which does not aim to explore social media's role as a mobilizing force, but rather seeks to examine social media as a source of gauging public opinion in an authoritarian state (see Olesen, 2011).

---

[7] Netizens often use creative euphemisms to evade censors, as in this instance, where "gate" (*men*) was added to the end of a term to evoke the Watergate scandal.

Table 1.1
Time Line of Shanghai Husi Incident and Other Food Safety Concerns, July–August 2014

Date	Event
July 20	Dragon TV airs undercover footage of food-safety violations at Shanghai Husi Group (上海福喜集团), including the reprocessing and repackaging of expired meat products subsequently supplied to McDonald's and KFC in China
July 22	Shanghai regulators seize 100 tons of expired meat from Shanghai Husi
August 3	Six senior Shanghai Husi executives detained by Shanghai police
August 6	China's FDA publishes new food-safety draft document on recalls, clarifying that food producers and businesses—not the government—bear primary responsibility for ensuring food safety
August 9	Walmart accused of food-safety violations in Chinese branches
August 11	FDA requires five foreign restaurants—including McDonald's and KFC—to make supplier information accessible to the public
August 12	Shanghai lawmakers draft food-safety traceability system proposal
August 18	Heinz recalls baby cereal in China due to high lead content
August 29	Shanghai judiciary officially approves the arrest of six senior Shanghai Husi executives

CHAPTER TWO

Automated Analysis of Social Media Use in China

Seeking to understand sentiment about local government, national government, and Western companies as expressed over microblogging platforms by Chinese-language users, we explored online discussions about food safety and the Husi incident. We collected data from Twitter and Sina Weibo, obtained psycholinguistic indicators of the social media messages, and interpreted the quantitative output.

Analysis of this social media data was intended to address two main questions:

- What topics were Chinese social media users discussing related to recent food safety incidents?
- How did sentiment compare when Chinese social media users discussed local Chinese government, national Chinese government, or Western companies, such as those involved in the food-safety scandals? To whom did they assign blame or responsibility for the scandals?

Analyses included, for example, plotting trends in the quantitative indicators against a time line of food-safety events in China to determine whether they coincided with these events. To analyze the data over time, we aggregated the Chinese-language microblog posts by week, which seemed most appropriate for broad trends over a several-month period (from the beginning of June 2014 to the end of August 2014). In other words, we combined social media texts for each one-week period to produce a single set of quantitative indicators for that week. We then tracked these indicators over the time period following the widely discussed food-safety scandal and examined whether patterns in these indicators coincided with specific events related to the scandal, to gain insight into how people may have felt before, during, and after those events.

We also analyzed the data by comparing the quantitative indicators when social media users discussed specific topics. These topics were based on a taxonomy of search terms, a set of keywords we created that included issues relevant for food safety and the Husi incident. Other analyses involved comparing between subsets of tweets discussing the West (e.g., U.S.A., McDonald's), the local Chinese government, or the national Chinese government. (See the Appendix for the taxonomy we used to assign tweets to categories.)

Our Psycholinguistic Approach to Social Media Analysis

Psycholinguistic analysis of social media text is based on the idea that language use can reveal insights about internal psychological states—that is, thoughts and feelings such as positive or negative sentiment about a topic, or connectedness with a group or public figure. Our automated methodology developed quantitative indicators of these sentiments, focusing on examining the rates at which people posting on social media about food safety in China used certain categories of words. To interpret the attitude or emotion conveyed by these categories, we used established precedent from literature about the software tool used in our method: Linguistic Inquiry and Word Count (LIWC, pronounced "Luke") 2007 (Linguistic Inquiry and Word Count: The LIWC2007 Application, 2007; Pennebaker et al., 2007). This software allows for content analyses of written text, such as messages posted using social media. LIWC compares the words that appear in a given text against a predefined dictionary containing words grouped into linguistic (e.g., pronouns, articles) and psychological (e.g., emotion, cognitive) categories. These comparisons are used to calculate ratios indicating the percentage of total words in a text that fall within each category. These ratios represent the relative frequency with which each word category is used in the text, which provides insight into the writer's attitudes, emotion, intentions, etc.

Extensive research has shown certain categories of words to be particularly meaningful. For instance, although pronouns represent a small proportion of overall word use, they actually reveal a considerable amount about an individual's psychological state (Pennebaker and Chung, 2007). By performing a qualitative analysis of the quantitative results generated by LIWC—that is, exploring how people's usage of these word categories changes over time and interpreting those changes in a real-life context—one can draw conclusions about public opinion and mood.

Certain categories of word usage may be particularly useful in understanding attitudes and opinions as expressed over social media:

- *Pronouns.* Usage of these linguistic function words indicates the focus of the author's attention. First-person singular pronouns (e.g., I, me, my) suggest self-focus, which is associated with negative states such as threat, depression, or insecurity. First-person plural pronouns (e.g., we, us, our) suggest that the author views him or herself as part of a larger group, implying a sense of shared group identity or community. Third-person pronouns (e.g., he, she, they) suggest the author is describing a group of which he or she is not a member, implying a different identity from that of the group (i.e., "othering").
- *Positive and negative emotion words.* Using more of these words may suggest that the author feels more positively or negatively about the topic of discussion.
- *Moral/legal responsibility words.* Usage of these words in discussion of a particular entity (e.g., local Chinese government, national Chinese government, Western companies involved with food) may suggest that authors associate responsibility for food safety with those entities.

For more detail on these word categories used in our analysis, and how they may be further used in analyses of social media, see Elson et al. (2012).

We Used C-LIWC to Analyze Chinese Language Text

Chinese LIWC (C-LIWC) is based on the translation of the English LIWC 2007 dictionary into traditional characters (Huang et al., 2012). The developers, Taiwan-based researchers, used a lexicon from Academia Sinica to develop their version of LIWC. Because LIWC was originally designed for the English language, conducting linguistic research on Chinese texts involves a number of potential challenges. First, there are a number of Chinese phrases, idioms, and slang words that are not terms that English speakers would use and thus difficult to translate into English. These terms would presumably not be in the English LIWC dictionary and would need to be added to the Chinese LIWC dictionary. Indeed, C-LIWC contains several unique categories, including second-person plural pronouns and several tense markers. Second, much of the psychological research on which LIWC is based was conducted in Western contexts and therefore may not be applicable to the Chinese population. Li and colleagues (2012) compared the outputs from English and Chinese versions of LIWC, describing similarities and differences between the word categories captured and analyzed by each version. Combining multiple LIWC categories into larger components, they found several psycholinguistic features common to each language, including several pronoun types (including those relevant for this research, such as first-person singular, first-person plural, third-person plural), negativity, narration, and social processes. Other psycholinguistic categories (e.g., sexual, social relations, and achievement), the authors argued, were more affected by cultural differences.

Collecting and Preparing the Sample of Chinese Social Media Data

We first selected samples of microblog posts made by Chinese-language users that discussed either the overall topic of food safety or the Husi incident in particular. We originally planned to exclusively analyze data from Sina Weibo, the most widely used microblogging platform in China. Yet unforeseen constraints with our social media data provider, DataSift, prevented access to historical data, limiting us to Weibo data from the time of data collection (September 2014) rather than capturing social media discussion immediately following the Husi incident in July 2014. Moreover, most sensitive posts on Weibo not automatically filtered (based on the presence of certain trigger words) are manually removed by Chinese censors within 24 hours of being posted. At the time of data collection, DataSift did not have a way of ensuring that Weibo messages are captured before any such removals. However, given that the discussion surrounding food safety on microblogging platforms in the wake of the Husi scandal was unlikely to contain many posts with collective action potential and more likely to consist of individual responses commenting on the event, this data collection method should have captured most of the Weibo posts made on the subject of food safety (see King, Pan, and Roberts, 2013, and King, Pan, and Roberts, 2014).

To supplement the limited Weibo data, we included a sample of Twitter data from the originally planned time period: June to August 2014.[1] Because some mainland Chinese dissidents circumvent China's "Great Firewall" to actively use Twitter to communicate with their followers in mainland China and in the West, we sought to compare what social media users who could include these (presumably) elite dissidents had to say about the Husi scandal with what might be found in a larger pool of public opinion from Weibo.[2] We note, however, that in the interest of consistency, we applied similar data collection, aggregation, and processing steps to both Twitter and Weibo samples.

Defining the Twitter and Weibo Samples

We collected samples from Twitter and Weibo by querying the time periods mentioned above for tweets or weibos that either used the terms *food safety* (食品安全) or *Husi incident* (福喜事件). First, we created a taxonomy of search terms relevant to either food safety or the Husi

---

[1] Data-collection limitations precluded collection of Twitter data from September 2014, preventing comparison with Weibo data that could establish whether Twitter data were representative of missing Weibo data.

[2] The small volume of geotagged tweets meant that we could not be certain that the tweets came from China or that they reflected the attitudes of Chinese netizens.

Table 2.1
Number of Food Safety Microblog Posts by Category

Category	Total Number of Tweets June 1–August 31, 2014	Total Number of Weibos August 26–September 8, 2014
Local government	404	248
National government	1,211	270
U.S. companies	569	19
Total	2,184	537

incident. We categorized these search terms according to various topics and actors, such as whether they refer to the local or national Chinese government, or other potentially relevant actors, such as Western fast food companies. Table 2.1 shows the number of tweets or weibos for each of the search terms.

We categorized the food-safety tweets according to whether they contained references to local government, national government, and U.S. companies (Table 2.1). Notably, there were a significant number of tweets that used the word for America (美国). We categorized tweets according to "U.S. companies" in order to gauge opinion toward the United States or its companies.

Table 2.1 shows weibos that discussed food safety collected during two weeks in August and September 2014. These were also categorized according to whether they made reference to local government, national government, and U.S. companies. Overall, however, there were relatively few microblog posts in a given category, which potentially limits the extent to which interpretations of these posts may be generalized.

Cleaning and Aggregating the Data

Prior to analysis, the raw data must be processed to ensure that it is in the suitable format. Given text-limit constraints and other peculiarities, social media data may be easier to analyze if certain data-cleaning steps are performed (for examples, see Elson et al., 2012). For this exploratory study, we performed minimal data cleaning on the Chinese-language text. One important step involved the particular search terms we used. We discovered that the simplified Chinese LIWC dictionary categorizes the word *safety* ("安全") as conveying positive emotion. This would be problematic for our analysis because when the phrase *food safety* ("食品安全") appears in the text—which it did frequently, as that was one of our primary queries—it is fragmented into "食品" and "安全," with "安全" being categorized as a positive emotion word. This appeared to skew the ratios of positive to negative emotion words in the sample (thereby suggesting that the tweets were more positive than they actually were), so we created a new dictionary that removed "安全" from the "positive emotion" category. Comparing LIWC analysis for a sample of Husi incident tweets from, for instance, the week of July 20–26, the simplified Chinese LIWC (SC-LIWC) dictionary categorized 1.14 percent of the words as conveying positive emotion, while the modified dictionary with the term "安全" removed from the positive-emotion category categorized 0.49 percent of the words as conveying positive emotion.

Next, the data were aggregated to the appropriate level of analysis. Given the time line (a few months) and the volume of social media messages we collected, we chose to aggregate the data by week.

Text Segmentation

Chinese text is written without spaces between words. Before conducting linguistic processing, we must add spaces between words so that C-LIWC can recognize and categorize words. An extensive body of literature exists on the challenges associated with segmenting Chinese into words (including the question of what exactly constitutes a *word* in modern Chinese) (Wong et al., 2010).

We segmented the social media text (i.e., tweets/weibos aggregated by week) using the Stanford Word Segmenter (2014), as recommended by creators of Simplified Chinese LIWC (Huang et al., 2012). The Stanford Word Segmenter comes with two preprogrammed segmentation standards: Chinese Penn Treebank (Xue et al., 2013) and Peking University (Duan et al., 2003). We used the Chinese Penn Treebank corpus, which contains over 1.5 million words and is based on "Chinese newswire, government documents, magazine articles, various broadcast news and broadcast conversation programs, web newsgroups, and weblogs" (Xue et al., 2013).

Processing and Interpreting the Data in Context of Chinese Social Media

We processed the segmented texts with C-LIWC and conducted qualitative interpretations of the quantitative C-LIWC output. Because most people writing in Chinese online appear to use simplified Chinese, we used the version of C-LIWC based on the version containing a dictionary of simplified Chinese characters (SC-LIWC). The creators of SC-LIWC, researchers at Academia Sinica, have found a tag rate of over 80 percent in a sample of 30 text files about breakups and part-time jobs collected online (Huang et al., 2012).

In their analysis of microblog posts, however, researchers at the University of the Chinese Academy of Sciences found that SC-LIWC alone categorized approximately 40 percent to 50 percent of words (Gao et al., 2013). Seeking to improve on this categorization rate, these researchers created a separate dictionary based on the 5,000 words most frequently used on Sina Weibo that were not already included in the SC-LIWC dictionary (Gao et al., 2013). With the addition of this supplemental dictionary, SC-LIWC was able to categorize a greater proportion of microblog posts: approximately 50 percent to 60 percent of words. We found similar results in our sample—for example, in the week of July 20–26 of our Husi incident sample, the SC-LIWC dictionary categorized 40.9 percent of words, while the SC-LIWC dictionary supplemented with the microblog-specific dictionary categorized 54.3 percent of words.

The categorization by SC-LIWC may appear to be a small proportion of the total words in our sample. As a comparison, LIWC2007 (i.e., English-language LIWC) captures, on average, 86 percent of words used in writing or speech (LIWC.com). Because people write and speak differently according to communications medium, microblogging texts may simply contain fewer words categorized by the LIWC and SC-LIWC dictionaries, which were based on more-formal or official texts (e.g., textbooks, news articles). This should not, however, be considered missing data. Rather, it may be that people writing over social media—and, in particular, in the Chinese language—simply use smaller proportions of the words that fit these predefined categories and which have been shown to be psychologically meaningful.

After analyzing the social media messages from Twitter and Weibo using C-LIWC, we plotted the trends against a time line we constructed of key events during the relevant time period. We also compared the C-LIWC indicators according to whether the social media messages discussed local Chinese government, national Chinese government, or Western companies.

Potentially complicating our interpretation of Chinese public opinion, there were a number of duplicated messages on both Twitter and Weibo that appeared to have been posted either by the Shanghai municipal government or by local news outlets about the Husi incident. In some cases, the duplicate posts reflect retweets, which we did not filter out, assuming that the content reflected the opinion of those who retweeted them. In other words, these duplications are meaningful in that they imply that the person or entity who took the effort to retweet or post a news article considered it important and likely at least somewhat representative of his or her views.

CHAPTER THREE

Results

Being limited to Weibo data current at the time of collection, we did not find much discussion on Chinese domestic microblogs related to the Husi incident. This was to be expected, given that the original news report that led to the food-processing scandal had occurred a few months prior to data collection, with plenty of time for discussion to die down. To supplement the findings from this small data set, we ran identical queries on the Twitter data from the time of the Husi incident and compared the two data sets. To understand attitudes, including sentiment and attributions of blame or responsibility, we analyzed discussion on Twitter and Weibo using the same sets of C-LIWC indicators: frequency of usage of pronouns, emotion words, and specific terms referring to moral or legal responsibility.

Tweets About Food Safety and Husi Incident: Trends over Time

First, we explored sentiment expressed over Twitter relating to general discussion about food safety. Changes in post–Husi incident sentiment about food safety suggest that Chinese public opinion is closely attuned to both local and national government responses to food-safety crises. The trends shown in Figure 3.1 indicate that, during the week when the Husi incident occurred, negative sentiment sharply increased, while positive sentiment decreased. (The negative-sentiment word category is the aggregate of the word categories for anxiety, anger, and sadness.) Expressions of negative sentiment remained elevated—that is,

Figure 3.1
Emotion Word Usage in Tweets About Food Safety

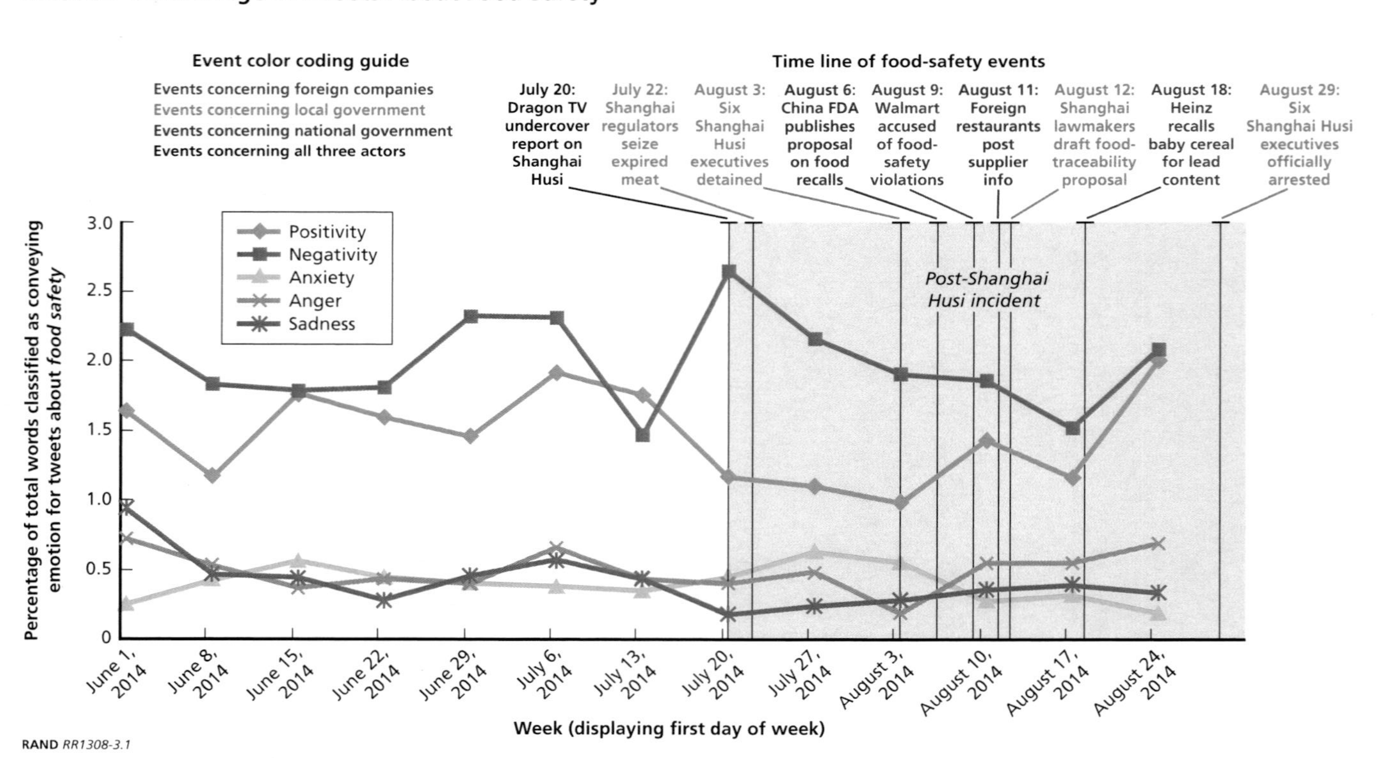

people wrote more negatively—through the week of August 3rd, when Shanghai police announced that they had detained six executives of Shanghai Husi. By August 11th, when the national government asked foreign fast-food chains such as McDonald's and KFC to post supplier information on their Chinese websites, negative sentiment when discussing food safety had returned roughly to pre-incident levels.

In the immediate aftermath of the Husi food-safety scandal, Chinese netizens appeared to use high rates of first-person singular pronouns in their Twitter posts specifically about the Husi incident (Figure 3.2). This suggests an initial reaction of negativity or depression to the news of the scandal, a pattern that is also reflected in the trends in sentiment when discussing food safety generally (Figure 3.1) and the Husi incident specifically (Figure 3.3). By the week after the scandal broke, however, netizens' usage of first-person plural pronouns increased, suggesting that netizens may have felt some sense of community or group cohesion in their shared suffering and frustration about the incident. By August 3rd, when Shanghai police arrested the six Shanghai Husi executives held responsible for their failure to assure safe practices in their company's meat production, netizens' use of third-person plural pronouns peaked, perhaps suggesting an "othering," or distancing, from those executives.

This dual pattern of first-person pronoun usage (i.e., increases in both first-person singular and first-person plural pronouns) reflects similar patterns seen in certain other large-scale, traumatic incidents. For instance, in tweets about the major protests immediately following the 2009 presidential election in Iran (Elson et al., 2012), and in newspaper coverage of a bonfire accident at Texas A&M (Gortner and Pennebaker, 2003), first-person singular pronoun usage spiked immediately after the respective traumatic event, reflecting a self-focus that may suggest intense negative or depressive states, as would be expected. In both cases, first-person plural pronoun usage—which signifies a desire for interaction with others—also increased following the traumatic event, but at different points in time. The extent and variation of collective focus may reflect differential coping and the amount of social cohesion among the participants in the discussion.

Figure 3.2
First-Person Pronoun Usage in Tweets About Husi Incident

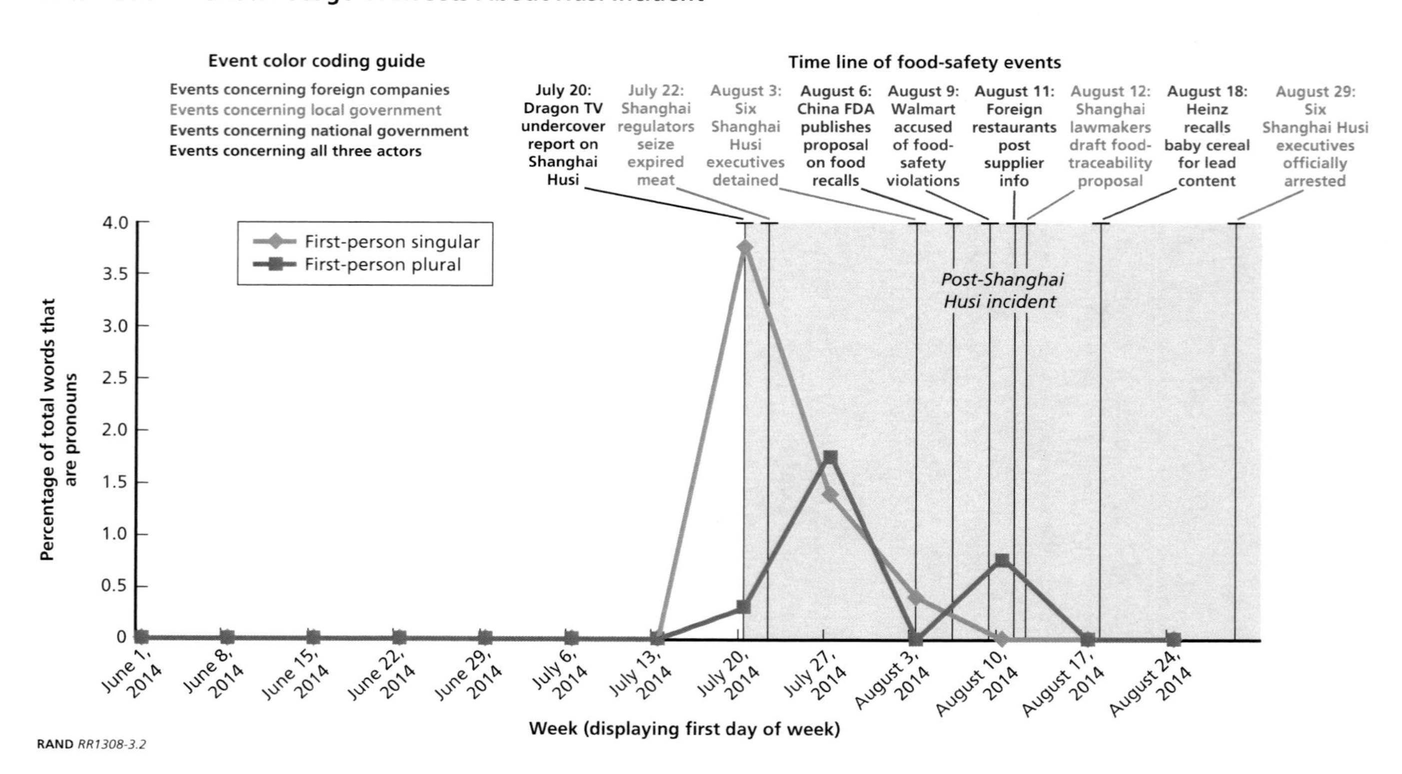

Figure 3.3
Other Pronoun Usage in Tweets About Husi Incident

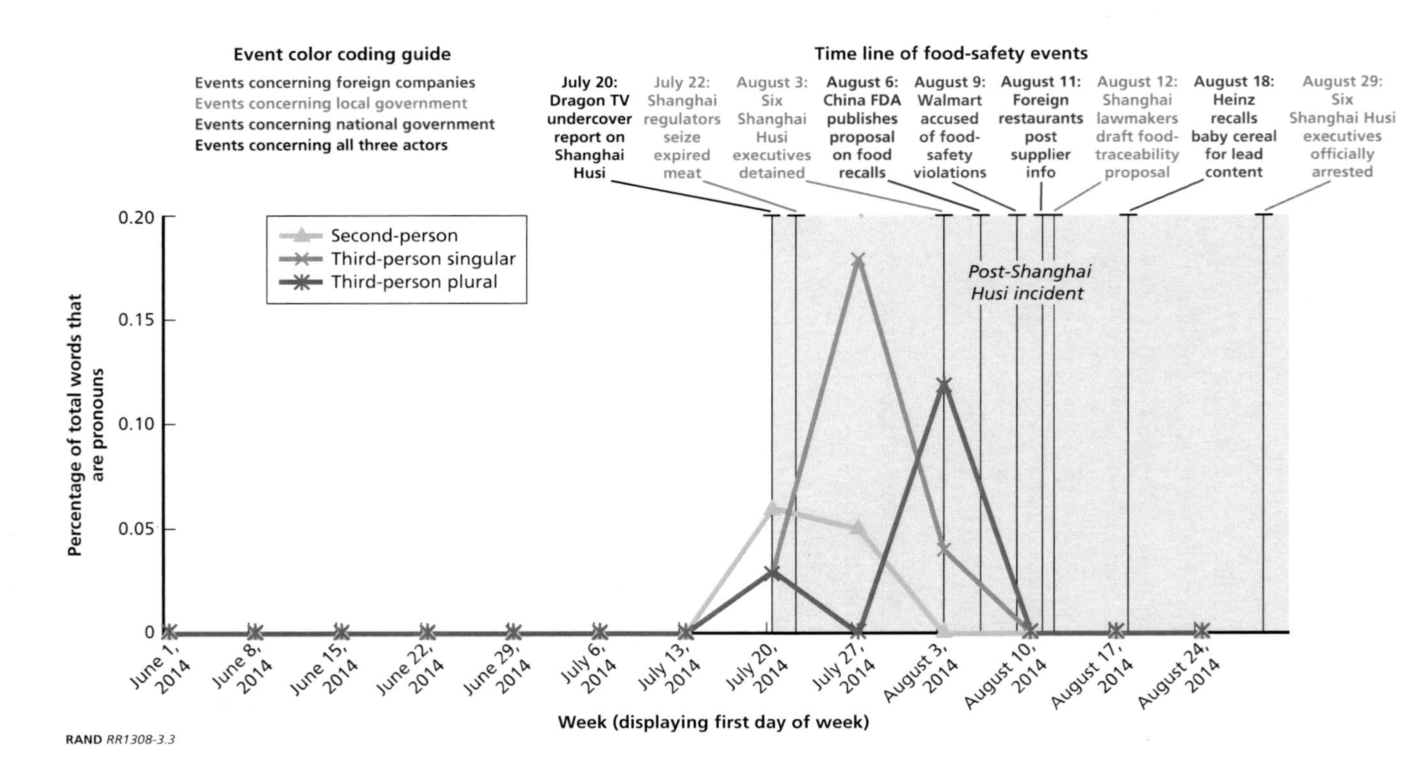

As shown in Figure 3.4, sentiment in tweets specifically discussing the Husi incident was much more negative than in posts about food safety after the scandal broke on July 20th. In the four weeks following the crisis, negative sentiment remained relatively constant, in spite of actions undertaken by both national and local government during this time period. Inferring the target of the negative sentiment expressed in these posts should be done with caution (and likely requires careful reading of the posts)—the sustained elevated negative sentiment could have been directed either toward U.S. companies such as McDonald's and KFC, or toward the Chinese government. In this case, the spike in negative sentiment during the last week of August likely reflects netizens' negative feelings toward the six Shanghai Husi executives, whose official arrest after detainment was announced on August 29.

Discussions About Chinese Government and U.S. Companies in Food-Safety Tweets

Netizens' frustration regarding the incident may have been directed less toward the local government and more toward the national government and U.S. companies involved in the scandal. Of note is the "othering," or distancing, from both the national government and U.S. companies suggested by netizens' prominent, frequent use of third-person plural pronouns (Figure 3.5). Third-person plural pronoun usage often reflects references to opposition groups or governments (Pennebaker and Chung, 2008), which in this case may imply that netizens viewed the national Chinese government and U.S. companies as the opposition or the problem in food safety.

Perhaps echoing this view of the national government and U.S. companies as the opposition, netizens used about twice as many negative emotion words regarding the national government and U.S. companies as regarding the local government (Figure 3.6). In particular, netizens appeared to direct their feelings of anger toward the national government rather than the local government; tweets containing references to the nation and to the national government contained over four times as many words conveying anger as tweets containing references to municipal- or provincial-level authorities. Similarly, tweets about the

Figure 3.4
Emotion Word Usage in Tweets About Husi Incident

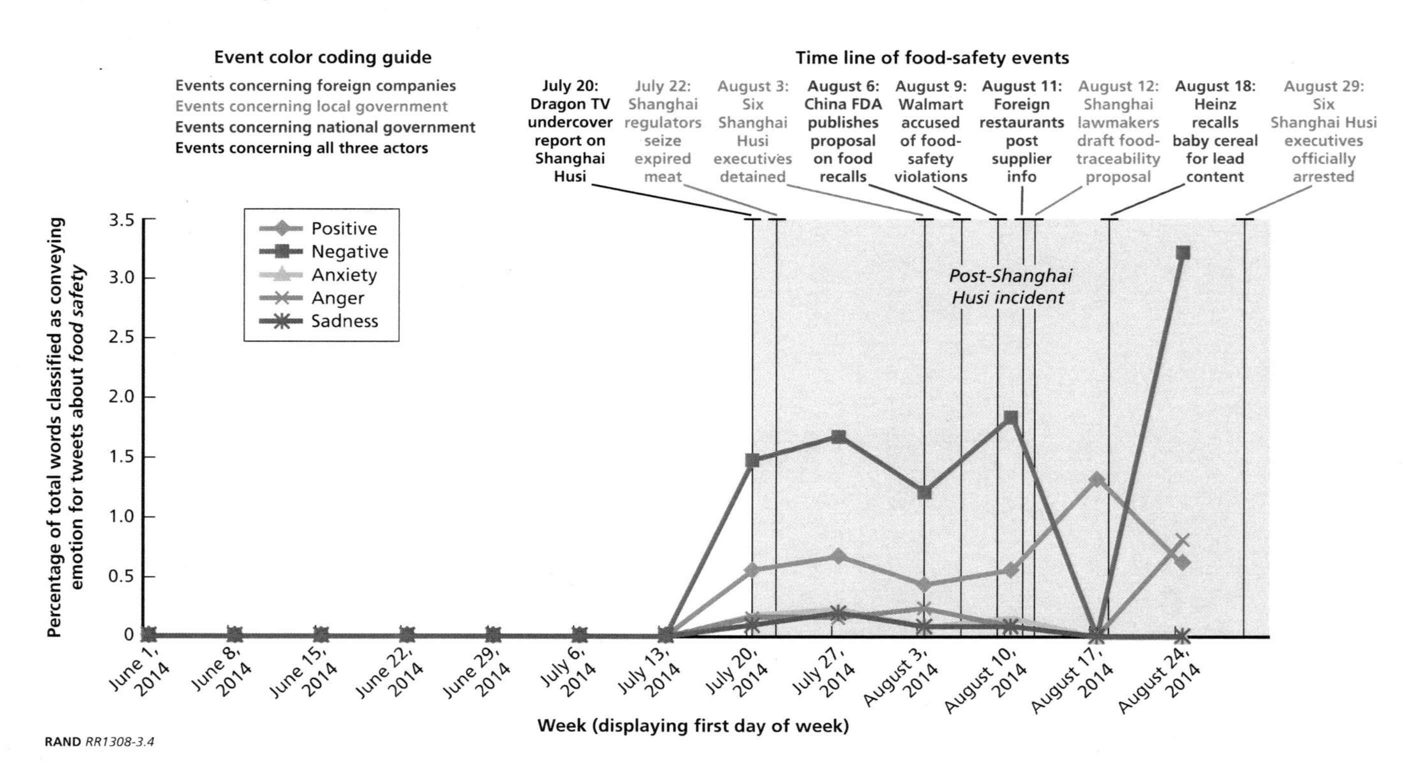

Figure 3.5
Comparison of Pronoun Usage in Tweets About Food Safety Referencing Local and National Government and U.S. Companies

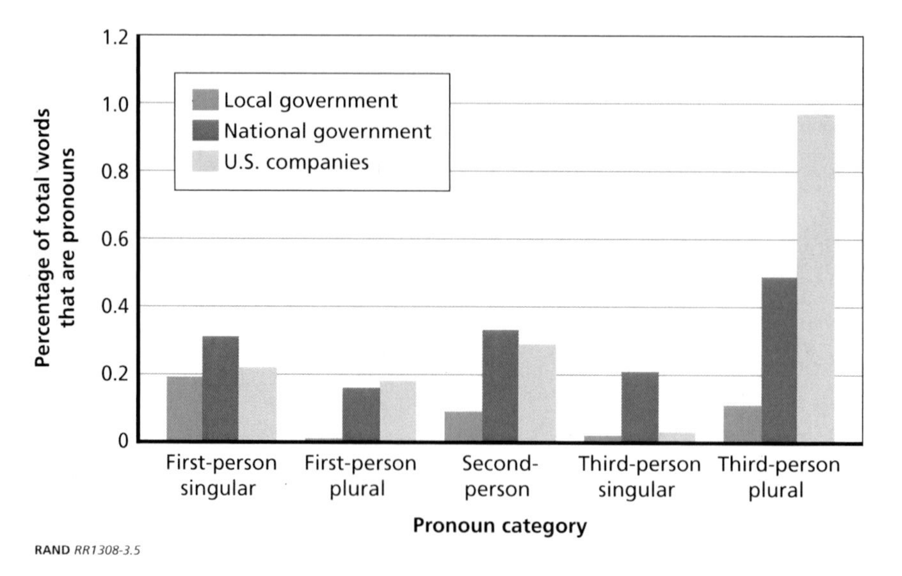

RAND *RR1308-3.5*

Figure 3.6
Comparison of Emotion Word Usage in Tweets About Food Safety Referencing Local and National Government and U.S. Companies

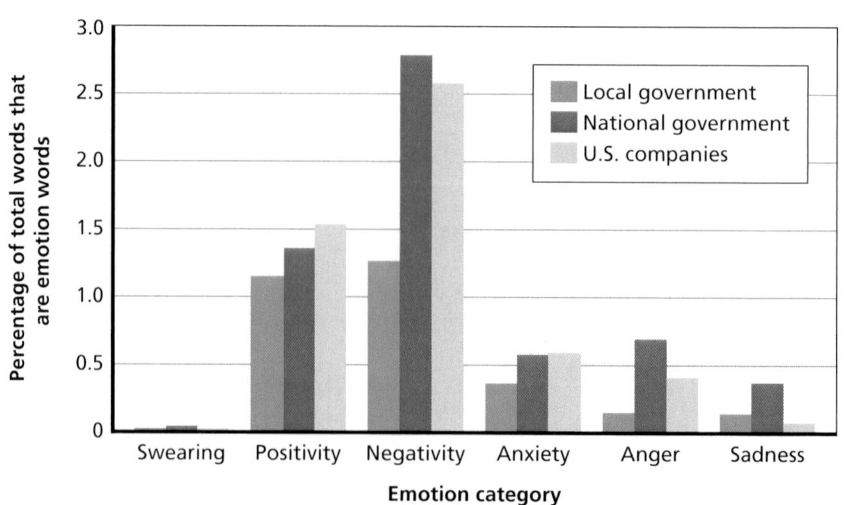

RAND *RR1308-3.6*

national government contained more than twice as many words conveying sadness as tweets about the local government.

As a way of exploring where netizens placed blame for food safety issues, we examined their usage of several legal and moral terms (Figure 3.7). While an overall picture may be unclear, it appears that local governments were viewed as being responsible for (or perhaps particularly effective in) punishing those responsible for their illegal behavior. Also, the term for corruption appeared almost twice as often in tweets about the national government as in tweets about the local government. This may reflect either popular attitudes about corruption at the national and local levels or merely that people spoke frequently about corruption in China or the whole country in general.

Further manual content analysis may be needed to disentangle specific interpretations of these results. For instance, a limited reading of the tweets suggests an explanation for why there appeared to be so much criticism directed at the national government—a fair proportion of the tweets seemed to have been posted by news outlets. This could explain—or perhaps complicate—these findings, as tweets and

Figure 3.7
Comparison of Legal and Moral Word Usage in Tweets About Food Safety Referencing Local and National Government and U.S. Companies

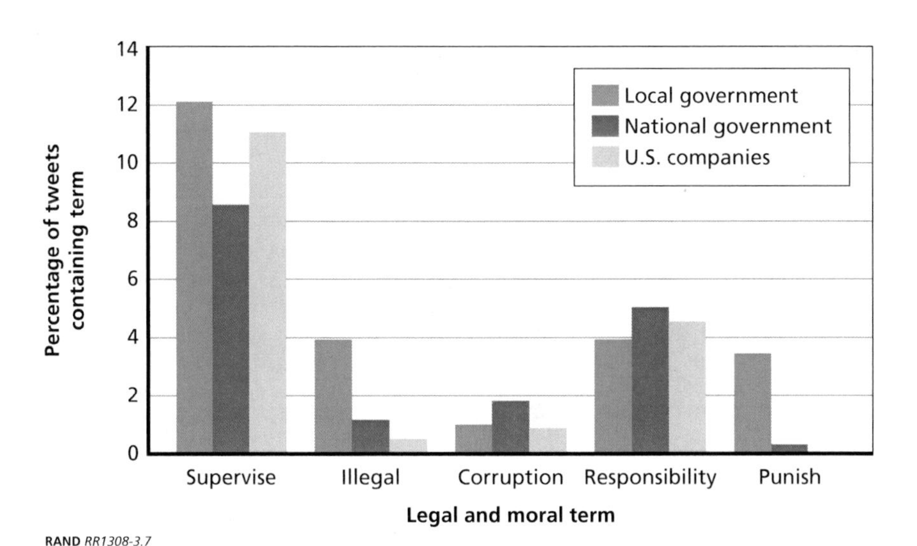

RAND *RR1308-3.7*

articles from Western news outlets (e.g., *New York Times, Wall Street Journal, Guardian*) in Chinese may skew more critical about the Chinese national government than posts made by Chinese netizens.

Discussions About Chinese Government and U.S. Companies in Food Safety Weibos

Given the uncertainties associated with the Twitter sample, including the possibility that only highly motivated and perhaps biased Twitter users would have surmounted the firewall, we also examined attitudes toward local government, national government, and U.S. companies using Weibo data.

Analysis of pronoun usage on Weibo suggested that people felt a greater sense of community and less depression or insecurity regarding the national government than in reference to the local government (Figure 3.8). Interestingly, these attitudes toward the local and national government appear to contradict those suggested by the Twitter analy-

Figure 3.8
Comparison of Pronoun Usage in Weibos About Food Safety Referencing Local and National Government and U.S. Companies, August 26–September 8, 2014

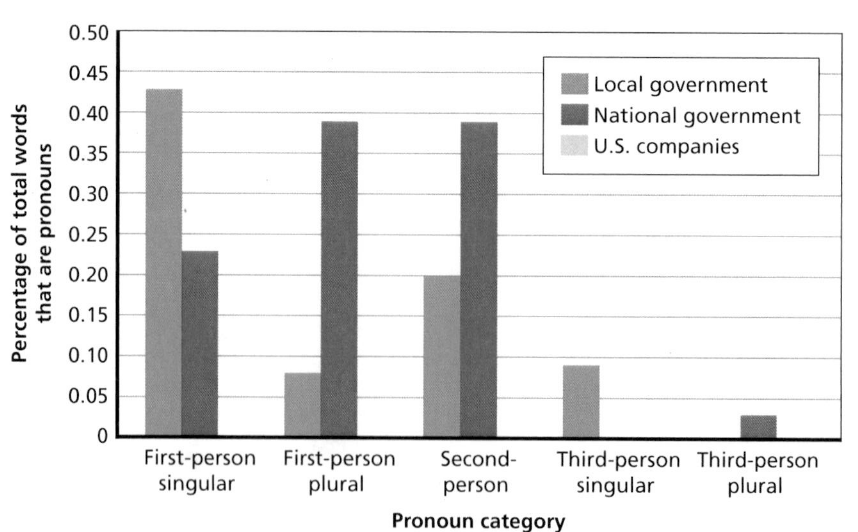

RAND *RR1308-3.8*

sis—Weibo netizens seemed to regard the national government more positively and the local government more negatively in discussing food safety, while Twitter netizens expressed the opposite sentiments. It is important to note, however, that the Twitter data and Weibo data come from two separate time periods—the Twitter data were collected before, during, and after the food safety scandal (from the beginning of June to the end of August 2014), while the Weibo data were collected during two weeks at the end of August and beginning of September 2014. As a result, it was not possible to directly compare data from both sources collected over the same time period.

The percentage of first-person singular pronouns in weibos about the local government was almost double that in weibos about the national government, suggesting that netizens felt more upset when discussing food safety and the local government on Weibo than when discussing food safety and the national government. It is possible but unlikely that Weibo posts about the national government were censored. First, posts about food safety, in this instance, are unlikely to involve calls for collective action or public protest, which constitute the vast majority of censored posts (see King, Pan, and Roberts, 2013 and 2014). Second, posts about food safety are unlikely to contain terms that would be censored immediately, based on keyword filters. We were fairly confident that our data collection captured posts not censored immediately, although DataSift was unable to confirm this. In contrast, the percentage of first-person plural pronouns in weibos about the nation or the national government was almost five times that of weibos about the local government, suggesting that netizens felt a greater sense of kinship in discussing food safety in China at large than in reference to the city or provincial levels.

This may reflect either censorship of Weibo (i.e., weibos critical of the national government's role in guaranteeing food safety and advocating concrete responses may be removed and thus not included in our analysis) or simply differences in populations using Weibo versus Twitter (i.e., there could be an increased presence of dissidents or overseas Chinese on Twitter versus on Weibo). Note that very few posts in our sample mentioned America or U.S. companies; this is likely a result of the time frame during which the posts were collected (during two weeks over a month after the food scandal broke).

Netizens on Weibo—in direct contrast to those on Twitter—wrote far more positively about the national government (or about *China* in general) than about the local government (Figure 3.9). However, netizens still appeared to direct their feelings of anger toward the national government rather than the local government; weibos containing references to the nation and to the national government contained over four times as many words conveying anger as weibos containing references to municipal or provincial level authorities. Similarly, weibos about the national government contained over three times as many words conveying sadness as weibos about the local government.

Weibos containing terms referring to America in general or U.S. companies such as McDonald's and KFC specifically contained far

Figure 3.9
Comparison of Emotion Word Usage in Weibos About Food Safety Referencing Local and National Government and U.S. Companies, August 26–September 8, 2014

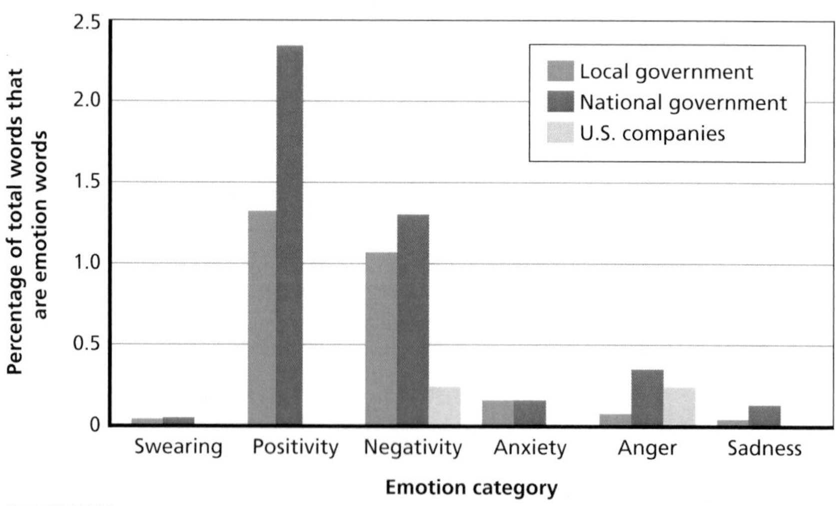

RAND *RR1308-3.9*

Figure 3.10
Comparison of Legal and Moral Word Usage in Weibos About Food Safety Referencing Local and National Government and U.S. Companies, August 26–September 8, 2014

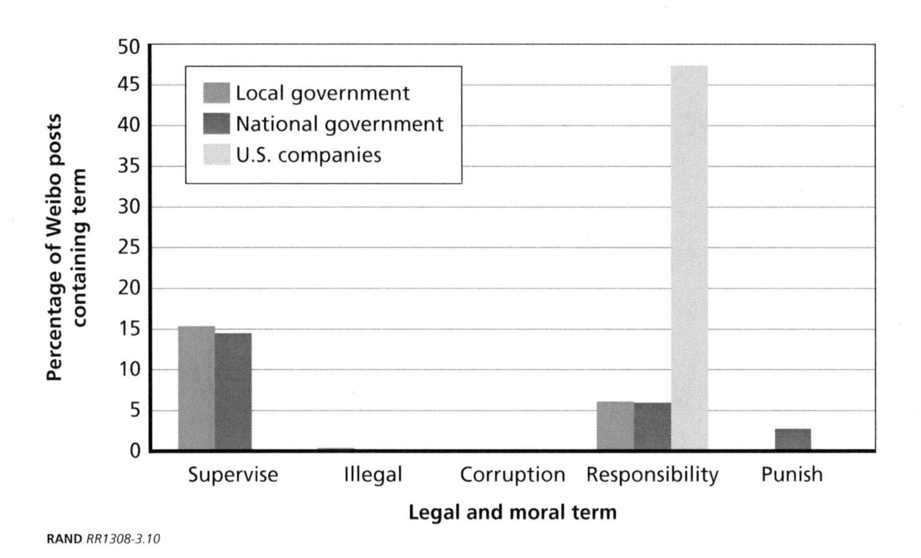

RAND *RR1308-3.10*

more posts about responsibility or blame (Figure 3.10). Unfortunately, without reading a great many more posts, it is difficult to know whether this suggests approval of the United States and its handling of food safety relative to China, or whether this suggests that netizens blame or hold the United States responsible for this food safety scandal.

Manual content analysis of weibos discussing the Husi incident turned up posts from quite a few news outlets and official organizations, as well as posts from individuals, including these examples:

> Supervision of food safety is not just the government's responsibility. It also requires meetings among relevant industries, organizations actively promoting corporate self-discipline, and actively troubleshooting and inspecting problems one-by-one in a timely manner. At the same time, everyone has a responsibility to supervise society. If every person had a greater desire to pay more atten-

tion to "things that don't concern them" and report [problems] in a timely fashion, then food could possibly become a little bit safer.

In light of the Chinese food market's current situation and national condition, China's food safety problems require severe punishment and drastic penalties, otherwise various types of "food safety" incidents will be difficult to catch and stop.

CHAPTER FOUR

Conclusions

The research described in this report explored how and whether Chinese-language online discussion could reveal useful insights about Chinese public opinion. Using food safety as a case study, our analysis of Chinese social media yielded findings with policy and methodological implications for understanding public attitudes regarding the Chinese government and which may inform future analyses of Chinese social media. In particular, applying our computerized methodology to gauge public sentiment expressed over social media suggested that the nature of Chinese-language online discussion about domestic politics varies by level of government (i.e., local versus national).

We explored discussions taking place on two social media platforms used in China: Twitter and Sina Weibo. Our Twitter data set was more complete, and therefore more relevant for analysis of the Husi incident, but may have represented a more specific set of social media users as compared with data from the more widely used Weibo.

Chinese Social Media Users Generally Expressed More Frustration with the National Government Than with Local Government, but Views Differed Across Twitter and Weibo

Indicators of sentiment and other attitudes expressed in Chinese-language discussions on microblogging platforms appeared very responsive to domestic political events related to food safety. In particular, Twitter users expressed more anger, sadness, and corruption-related words when discussing the national government than when discussing the local government. While Weibo users also expressed more anger and sadness when discussing the national government, they also used more positive emotion words. Moreover, Weibo users wrote with more second-person pronouns when discussing the national government, suggesting a feeling of kinship with China more broadly—perhaps a reflection of patriotism or nationalism. This overall increase in emotional expression regarding the national government may reflect a diversity of opinions among individual members of the public. Alternately, it may suggest conflicting feelings within the public or that topics of broad national interest generally stimulate interest and debate.

The divergences in expressed sentiment may also reflect different communities of users that hold different viewpoints and opinions. For instance, people tweeting in Chinese may be more politically active and critical of the national government, given the additional effort required to access Twitter. In contrast, users of Weibo—more widely used by far—could potentially be more representative of the overall population, at least in terms of political views. But it is difficult to tell whether the more positive impression of the national government portrayed on Weibo is genuine or colored by the national government's control over the Chinese media. The correspondence between public opinion on Weibo and the national government's interest in this case may reflect a successful public-relations framing campaign on the national government's part. It would be in the national government's interest to deflect governance scandals onto local government officials and Western countries or companies to divert criticism from itself, and this incident afforded the national government an opportunity to do exactly this. The fact that users still wrote in more emotional terms about the national government may suggest a number of possibilities. The first is that the national government failed to deflect this particular scandal onto either the local governments or the Western companies involved. Another explanation for this finding is that netizens may view food safety as a broad issue for which the national government has greater responsibility to legislate, regulate, and ensure. Alternatively, netizens might perceive the national government as having a greater responsibility to monitor and regulate foreign companies than the local government (since the companies implicated in the Husi scandal were suppliers of Western fast-food chains). Yet another explanation might be that netizens viewed the local Shanghai government (or even the Western fast-food companies) as taking effective measures to solve the crisis once it occurred, while the national government was perceived more negatively for failing to catch the problem earlier. A final possibility is that this pattern simply reflects how publics everywhere respond to crises: Overall sentiment on Twitter about the Husi incident generally followed similar patterns in other large-scale traumas or disasters (e.g., Elson et al., 2012; Gortner and Pennebaker, 2003). These tweets contained greater rates of both first-person singular and first-person plural pronouns, suggesting reactions of negativity about the incident as well as a sense of community or group cohesion in that shared frustration, respectively. Attitudes about the Husi incident, however, were only available from Twitter (Weibo data from this time period was unavailable); thus, we could not compare attitudes about the Husi incident as expressed on either Twitter or Weibo.

Analyzing Chinese Social Media Presented Unique Challenges

In this exploratory analysis of Chinese social media, we encountered several challenges that limit the generalizability of the findings. In particular, the extent to which our data were representative of Chinese social media users was constrained, given that we analyzed mainly Twitter data for Chinese domestic public opinion, rather than the more widely used platform, Sina Weibo. Access to data for Weibo through our data provider, DataSift, was limited to messages that happened to be sent at the time of data collection going forward. This meant that we were largely unable to track the trajectory of social media discussion around the Husi crisis from its beginning until the present. In contrast, full historical data for Twitter was available. However, the use of Twitter as a data source suffers from uncertainty about the intended audience of tweets. Because user location is difficult to determine and Twitter access is blocked in China, the intended audience of these tweets is likely not the Chinese domestic audience. If Twitter users had intended to communicate with non–Chinese speakers or with Westerners specifically, they would likely have used English rather than Chinese. These data-availability issues limited the extent to which we could compare findings from Twitter against those from Weibo. For instance, we could not directly compare public opinion at the time of the Husi incident. A lack of historical access to the larger volumes of postings on Weibo also constrained the overall sample size of relevant social media postings. Given the overall sparseness of the Twitter and Weibo data, aggregating the posts (i.e., by week) was necessary but could also have led to less–fine grained and accurate analysis, particularly at the beginning and ending weeks of the overall data set, each of which contained less than a full week of data. Finally, there were a few cases in which it appeared that certain social media postings had been replicated in the data set two or three times. While such repetition could skew the indicators—which rely on frequency counts to calculate relative word usage—these repetitions were sufficiently isolated that they did not appear to have substantially affected the results.

Yet even had access to this data been unimpeded, interpreting individual public opinion when using Chinese social media, and especially Weibo data, faces other challenges and uncertainties that limit its utility. For instance, population data on the number and demographics of Twitter users in China is speculative. Estimates of Chinese Twitter users range from the hundreds of thousands to tens of millions (see Chapter One). The demographics of these Twitter users may be skewed by the necessity of bypassing the Great Firewall, suggesting that Chinese citizens sufficiently motivated to access Twitter may also be those most dissatisfied with the regime. Specific demographic information, such as location, is particularly difficult to determine. Few Twitter users geotag their tweets—12 tweets in our sample were geotagged (out of 3,488 tweets total). Moreover, even if more tweets were geotagged, the nature of Twitter access in China means that users' tweets could appear to be coming from the location of the proxy server or VPN being used to bypass the firewall, rather than from within China. For instance, the content of some tweets we examined clearly indicated that the user was located in Taiwan (e.g., use of traditional Chinese or content referring to Taiwanese locations or events). Thus, the Twitter posts cannot definitively be said to originate from Chinese citizens, only that they represent Chinese language–based discussion.

This ambiguity regarding the users of Chinese social media platforms further extends to the purposes for which those platforms are used. For instance, some of the posted content may not reflect true public opinion—that is, the opinion of individual social media users. A fairly trivial example is that news organizations often post tweets or microblogs. The inclusion of these postings thus may not reflect the exact opinion of individuals—however, these posts can nonetheless illustrate public opinion when they are retweeted or quoted by individuals. Another possibility with more-serious methodological implications is that people working on behalf of the Chinese government publish positive tweets about the government's responses to events (Sonnad, 2014). Such "ghost accounts," created to manipulate public discussion and opinion, generate uncertainty about who is actually posting and the extent to which social media content reflects genuine public opinion. In our sample, certain messages did appear to be government posts looking to shape opinion. For example, the following post—which in retweeting a news article does not assign blame but calls for punishing companies—was repeated a number of times:

> Han Zheng: In Shanghai, no matter what, companies must be punished for violating the law. Shanghai Party Secretary Han Zheng presided over a special meeting this afternoon, listening to the reports investigating the "Husi Incident." Han Zheng stressed that food safety is no small matter, but rather a serious matter of people's health and safety, and government regulators must adhere to the "five most stringent . . . "

Future Studies Can Extend Investigation of Chinese Domestic Politics and Social Media Platforms and Address Challenges of Automated Analysis

Future studies of Chinese public opinion should continue to explore political expression regarding domestic politics at all levels of government and seek to address the challenges we have identified regarding automated analysis of Chinese-language social media. Regarding the issue of food safety, broadening the keywords used to filter discussion to include more generally used phrases (e.g., *meat quality* or *fast food*) could capture a wider range of discussions relevant for understanding attitudes about food safety and its regulation. In addition, future studies might consider exploring a more rigorous methodology for selecting the words used to categorize tweets according to references to local and national government; different terms might result in more-accurate classifications and thus more-accurate reflections of public opinion toward these different levels of government.

Many other domestic issues relevant to Chinese public opinion warrant further investigation. For instance, pollution (e.g., soil, water, air) and other environmental concerns have been widely reported and debated in the news media and online. Netizens may hold divergent views about national and local government on environmental issues. Though the national government is arguably responsible for instructing local officials to prioritize economic growth over environmental protection, netizens may view the matter differently. For instance, netizens may applaud the national government for issuing mandates to reduce water and air pollution and blame local governments for ignoring new regulations—a common practice, given that local officials frequently lack financial incentives to police state-owned companies who dump industrial chemicals into water supplies and emit toxic gases into the air (Schmitz, 2014; Minter, 2014). Even if the public blames local government for environmental degradation, growing numbers of local environmental protests—which increasingly occur in urban areas instead of rural communities—may challenge the authority of the national government (Xu, 2014). Thus, social media discussion about pollution (e.g., smog), in conjunction with mentions of particular cities (e.g., Beijing, Shanghai, Chengdu, Guangzhou, Tianjin), could shed light on local versus national tensions regarding air pollution. When such messages are combined with other data, such as air quality, emissions, or factory and industrial outputs, it may be possible to depict regional and population differences in sentiment, helping to pinpoint where unrest is most likely to occur (and spread). In lieu of specific crises or incidents to examine, such an approach may suffice to understand broad public opinion about air pollution or similar domestic topics.

Another widely discussed issue relates to the concerns expressed during Occupy Central, the Hong Kong prodemocracy movement against the central Chinese government's declaration that a pro-Beijing committee would approve candidates for Hong Kong's 2017 chief executive election. The resulting frustration and anger from the Hong Kong protesters may be construed as a desire for local government control as opposed to that of a central government. An analysis of sentiment expressed over Twitter regarding the leaders of the local and national governments (Leung Chun-ying and Xi, respectively) revealed that Leung appeared to be the primary target of discontent (Yeung and Cevallos, 2014). Moreover, protesters felt relatively disconnected from the geographically and culturally distant Xi. Further analysis of social media discussions related to Hong Kong could also seek to explore economic trade and immigration issues. Hong Kong's and China's trade have become increasingly interlinked since Hong Kong's 1997 handover from British control, but Hong Kong's importance to China as a source of capital and a trade hub has diminished as the Chinese economy has grown. Nevertheless, the question of which—Hong Kong or China—needs the other more is unsettled and is, understandably, a point of discomfort for Hong Kong residents (*Economist*, 2014; Guilford, 2014). Hong Kong residents could, for instance, hold their local government accountable for failing to maintain sufficient economic independence from the mainland. Attitudes about immigration from mainland China to Hong Kong may also be worth exploring, given that it potentially involves immigrants on both ends of the economic spectrum. For instance, on the high end, people may try to park money outside the mainland, while at the low end, people may be simply looking to find work. Either case may lead to resentment among Hong Kong residents, raising the question of whether they view local government as accountable for curtailing immigration or hold the national Chinese government responsible for stemming the flow from the other end.

Social media research of Chinese political issues may also yield insights used in unintended ways. For instance, information on how the Chinese public regards the government's policy actions could be used to guide further action, including social control (Creemers, 2015). Researchers of Chinese social media should consider the implications of how policymakers may employ their findings.

Finally, further research of Chinese social media that employs automated analysis methods should also seek to address some of the challenges described in this report. For instance, selection bias is a significant and recurring concern when attempting to draw conclusions about specific populations based on social media content. In China, many active Twitter users are members of the urban elite and may also be more likely to be dissidents. Also, Chinese Twitter users may post content at least partially geared toward foreign audiences, as opposed to mostly toward domestic ones. This manufactured content may thus be somewhat artificial, rather than revealing the population's true attitudes and sentiment. Understanding the extent to which elite-heavy platforms such as Twitter are representative of the broader public may require rigorous comparisons with some of the more widely used microblogging platforms, such as Weibo.[1]

Potential next steps along this path could include obtaining historical data from users in China (either by geotagging or by self-reported location in China) on platforms such as Twitter or Facebook, then comparing against a parallel Weibo or Tencent data set to potentially uncover any cross-platform differences. Such comparisons could, for example, allow for a more nuanced exploration of attitudes expressed at the time of the Husi incident. Collecting Twitter data for the month of September (up to the 16th) could allow com-

---

[1] Our study collected Twitter and Weibo data on just a few of the same days at the end of August 2014, yielding too little data to determine whether Twitter data might be representative of the missing Weibo data from July and August 2014.

parisons of the volume of conversations on Twitter and Weibo, making it possible to infer approximately how many posts Weibo might have had about food safety and the Husi incident at the peak of the crisis. It may also be useful to more directly compare political sentiment on emerging, rival social media platforms. Although Sina Weibo has been the dominant platform in China, there has been recent migration away from it. A random sample of 4,500 influential Weibo users (defined as those with over 50,000 followers) showed a 20 percent drop in aggregate monthly posts from January to August 2013 (Chin and Mozur, 2013). There may be several reasons for this decline in usage. First, the Chinese Communist Party (CCP) took steps that discouraged usage. In March 2012, the CCP required users to register their real names with their Weibo accounts. And in June 2012, the CCP began arresting hundreds of users posting "rumors" on Weibo. Research from the *Telegraph* suggests that Weibo posts "may have fallen by as much as 70 per cent in the wake of an aggressive campaign by the Communist party to intimidate influential users" (Moore, 2014). As a result of these actions, the total number of Sina Weibo mobile app active users decreased almost 11 percent to 67.5 million in August 2014 (China Internet Watch, 2014). Weibo users may also have abandoned the platform for its competitors, such as Weixin (also known as WeChat), whose more private interface allows people to share information with a small group of friends, with a smaller prospect of government censorship. Given the downward trends in Weibo usage, these other social media platforms may offer upward-trending usage that suggest their increasing importance and role in people's discussions and lives. Exploring these platforms may be fruitful in uncovering further insights and opinion, although the nature of Weixin's private platform may preclude aggregation of larger quantities of data for research. Future research could explore how to control for the social media content we identified as coming from municipal government, news outlets, or other organizations, given that they should not be assumed to reflect true public opinion.

More-extensive Chinese social media analyses than this exploratory effort could leverage more-sophisticated techniques to test the reliability and validity of these findings. For instance, establishing a control group, such as by collecting social media data from the pre-Husi time period, could help directly compare public opinion on Twitter versus Weibo or on other social media platforms. This additional data could also provide a useful baseline against which to compare subsequent shifts in word usage. Statistical tests (e.g., Kolmogorov–Smirnov, Granger causality) can help determine whether observed differences in word usage (e.g., pre- and postevent) were due to chance. Further, alternate ways of structuring data queries may lead to the collection of a different data set that could provide more-reliable assessments of Chinese sentiment. These could include, for example, collecting all geotagged tweets from China and tweets whose users self-report their location as within China or simply using more general and colloquial search terms—such as *fast food* or *meat* instead of the more formal *food safety*, as in the Husi case. Finally, exploring the number and characteristics of users who post about food safety may also provide additional insight into the approximate size of the online population discussing these topics. Techniques such as social-network analysis could identify particularly influential users or connections between users and characterize population segments with differing opinions or characteristics.

Concluding Thoughts

Psycholinguistic analysis of social media conversations may provide unique insight into public sentiment across global political contexts and not otherwise easily obtained. Discussions on Chinese microblogging platforms appear to offer a wealth of information to explore additional questions regarding public opinion on issues of domestic politics, although the value of such openly accessible platforms may be in decline as Chinese users respond to government measures aiming to curb the potential for social media to mobilize collective action. This initial exploration of Chinese social media about the issue of food safety offers an initial road map to conduct such analysis, including a methodology for analyzing Chinese social media, possibilities and challenges, required tools and steps, and comparisons with English-language social media.

APPENDIX

Search Taxonomy

This table lists the terms used as keywords to collect Chinese-language social media posts about several topics related to food safety, the Husi incident, and either local or national Chinese government. Numbers of tweets in each also are provided as a rough illustration of the relative amounts of discussion related to each of the terms.

Table A.1
Terms Used to Collect Chinese-Language Social Media Posts

Category (English Translation)	Chinese Term	June	July	August	Total Tweets
Total tweets (containing *Husi Incident* and/or *food safety*)	福喜事件 and/or 食品安全	708	1,648	1,132	3,488
Terms about the West					
USA	美国	48	133	73	254
McDonald's	麦当劳	0	339	99	438
Western fast food	洋快餐	0	118	8	126
KFC	肯德基	0	203	25	228

Table A.1—Continued

Category (English Translation)	Chinese Term	June	July	August	Total Tweets
Terms about the national government					
Government (NOT province NOT city)	政府	15	109	58	182
Country	国家	108	92	98	298
China	中国	251	533	385	1,169
Leader(s)	领导	2	3	3	8
CCP	中共	3	38	2	43
Whole country	全国	74	22	8	104
Terms about the local government					
Province	省	15	66	69	150
City	市	72	174	172	418
Han Zheng (CCP Committee Secretary of Shanghai)	韩正	0	32	0	32
	局长	3	3	73	79
Moral/legal terms					
Monitor and regulate/supervise	监管	17	135	81	233
Illegal	违法	16	44	20	80
Corruption	腐败	2	23	15	40
Responsibility/blame	责	20	122	50	192
Penalize	处罚	1	4	2	7
Severely punish	严惩	1	26	0	27

References

Bai, Jianrui, "2013 Weibo Users Development Report" ["2013 nian weibo yonghu fazhan baogao"], *Weibo Data Center*, January 1, 2014. As of November 26, 2014:
http://data.weibo.com/report/reportDetail?id=76

Balzano, John, "Three Things to Watch for in Chinese Food Safety Regulation in 2014," *Forbes*, February 5, 2014. As of May 19, 2015:
http://www.forbes.com/sites/johnbalzano/2014/02/05/three-things-to-watch-for-in-chinese-food-safety-regulation-in-2014/#2715e4857a0b31102cef51df

———, "Issues on the Horizon for Chinese Food Safety Law in 2015," *Forbes*, March 8, 2015. As of May 19, 2015:
http://www.forbes.com/sites/johnbalzano/2015/03/08/issues-on-the-horizon-for-chinese-food-safety-law-in-2015/#2715e4857a0b3c59f5df2541

Barboza, David, "Death Sentences in Chinese Milk Case," *New York Times*, January 22, 2009. As of Novemer 26, 2014:
http://www.nytimes.com/2009/01/23/world/asia/23milk.html

Barclay, Eliza, "Tainted Pork Is Latest Food Safety Scandal in China," NPR, April 29, 2011.

Chang, Pi-Chuan, Michel Galley, and Chris Manning, "Optimizing Chinese Word Segmentation for Machine Translation Performance," Association for Computational Linguistics, June 2008. As of May 19, 2015:
http://nlp.stanford.edu/pubs/acl-wmt08-cws.pdf

Chin, Josh, and Paul Mozur, "China Intensifies Social-Media Crackdown: Campaign Takes Toll on Public Debate, Popular Platform," *Wall Street Journal*, September 19, 2013. As of November 26, 2014:
http://online.wsj.com/articles/SB10001424127887324807704579082940411106988

China Internet Network Information Center, "The 33rd Statistical Report on Internet Development in China," January 2014. As of November 26, 2014:
http://www1.cnnic.cn/IDR/ReportDownloads/201404/U020140417607531610855.pdf

———, "The 35th Statistical Report on Internet Development in China," January 2015. As of October 9, 2015:
http://www1.cnnic.cn/IDR/ReportDownloads/201507/P020150720486421654597.pdf

Creemers, Rogier, "Cyber China: Upgrading Propaganda, Public Opinion Work, and Social Management for the 21st Century," *Journal of Contemporary China*, forthcoming, available via Social Science Research Network: December 2, 2015.

Duan, Huiming, Xiaojing Bai, Baobao Chang, and Shiwen Yu, "Chinese Word Segmentation at Peking University," 2003. As of May 19, 2015:
http://www.aclweb.org/anthology/W03-1722.pdf

Economist, "Why Hong Kong Remains Vital to China's Economy," September 30, 2014. As of January 16, 2015:
http://www.economist.com/blogs/economist-explains/2014/09/economist-explains-22

Elson, S. B., D. C. Yeung, P. Roshan, S. Bohandy, and A. Nader, *Using Social Media to Gauge Iranian Public Opinion and Mood After the 2009 Election*, Santa Monica, Calif.: RAND Corporation, TR-1161-RC, 2012. As of May 19, 2015:
http://www.rand.org/pubs/technical_reports/TR1161.html

Fewsmith, Joseph, *The Logic and Limits of Political Reform in China*, Cambridge, UK: Cambridge University Press, 2013.

Fu, King-wa, Chung-hong Chan, and Michael Chau, "Assessing Censorship on Microblogs in China: Discriminatory Keyword Analysis and Impact Evaluation of the 'Real-Name Registration Policy,'" *IEEE Internet Computing*, Vol. 17, No. 3, May/June 2013, pp. 42–50.

Funakoshi, Minami, "Yet Another Food Safety Scandal in China—Now Rat Meat Masquerades as Lamb," *Tea Leaf Nation*, May 5, 2013. As of November 26, 2014:
http://funakoshi1.rssing.com/browser.php?indx=8232254&last=1&item=9

Gao, Rui, Bibo Hao, He Li, Yusong Gao, and Tingshao Zhu, "Developing Simplified Chinese Psychological Linguistic Analysis Dictionary for Microblog," *Brain and Health Informatics*, Vol. 8211 of *Lecture Notes in Computer Science*, 2013, pp. 359–368.

Goodman, J. David, "As China Reins in Microblogs, Dissidents Find Haven on Twitter," *New York Times*, January 23, 2012. As of January 14, 2015:
http://thelede.blogs.nytimes.com/2012/01/23/as-china-reins-in-microblogs-dissidents-find-haven-on-twitter

Gortner, E. M., and J. W. Pennebaker, "The Archival Anatomy of a Disaster: Media Coverage and Community-wide Health Effects of the Texas A&M Bonfire Tragedy," *Journal of Social and Clinical Psychology*, Vol. 22, 2003, pp. 580–603.

Guilford, Gwynn, "Why China Doesn't Feel the Need to Back Down in Hong Kong," *Quartz*, September 29, 2014, As of January 16, 2015:
http://qz.com/272952/why-china-doesnt-feel-the-need-to-back-down-in-hong-kong

Huang, C. L., C. K. Chung, N. Hui, Y.C. Lin, Y. T. Seih, B. Lam, and J. W. Pennebaker, "The Development of the Chinese Linguistic Inquiry and Word Count Dictionary," *Chinese Journal of Psychology*, Vol. 54, No. 2, 2012, pp. 185–201.

Internet Live Stats, "Internet Users by Country (2014)," website, 2014. As of May 19, 2015:
http://www.internetlivestats.com/internet-users-by-country

Jiang, Jessie, "China's Rage Over Toxic Baby Milk," *TIME*, September 19, 2008. As of April 17, 2015:
http://content.time.com/time/world/article/0,8599,1842727,00.html

Jiang, Zhan, "Environmental Journalism in China," in Susan L. Shirk, ed., *Changing Media, Changing China*, Oxford, UK: Oxford University Press, 2010.

Johnson, Ian, "Do China's Village Protests Help the Regime?" *New York Review of Books Blog*, December 22, 2011. As of May 19, 2015:
http://www.nybooks.com/blogs/nyrblog/2011/dec/22/do-chinas-village-protests-help-regime

———, *Wild Grass: Three Portraits of Change in Modern China*, New York: Pantheon Books, 2004.

Kaiman, Jonathan, "China Arrests 900 in Fake Meat Scandal," *Guardian*, May 3, 2013. As of April 17, 2015:
http://www.theguardian.com/world/2013/may/03/china-arrests-fake-meat-scandal

King, Gary, Jennifer Pan, and Margaret E. Roberts, "How Censorship in China Allows Government Criticism but Silences Collective Expression," *American Political Science Review*, Vol. 107, No. 2, May 2013, pp. 1–18.

———, "Reverse-Engineering Censorship in China: Randomized Experimentation and Participant Observation," *Science*, Vol. 345, No. 6199, August 22, 2014, pp. 1–10.

Kuo, Lily, "KFC, Pizza Hut, and McDonald's Are Hit with a New China Food Scandal: Expired Meat Products," *Quartz*, July 20, 2014. As of November 26, 2014:
http://qz.com/237643/kfc-pizza-hut-and-mcdonalds-are-hit-with-a-new-china-food-scandal-involving-expired-meat-products

Li, Cheng, "Think National, Blame Local: Central-Provincial Dynamics in the Hu Era," *China Leadership Monitor*, Hoover Institution, Issue 17, January 30, 2006, pp. 1–24.

Li, Haiying, Zhiqiang Cai, Arthur C. Graesser, and Ying Duan, "A Comparative Study on English and Chinese Word Uses with LIWC," Association for the Advancement of Artificial Intelligence, 2012.

Linguistic Inquiry and Word Count: The LIWC2007 Application, website. As of September 10, 2015:
http://liwc.net/howliwcworks.php#index7

LIWC—*See* Linguistic Inquiry and Word Count.

Lorentzen, Peter L., "Regularizing Rioting: Permitting Public Protest in an Authoritarian Regime," *Quarterly Journal of Political Science*, Vol. 8, No. 2, April 2013, pp. 127–158.

Lu, Rachel, "Chinese Web Users Resort to Dark Humor to Mask Fears About Pig Carcasses in Shanghai River," *Tea Leaf Nation*, March 12, 2013. As of November 26, 2014:
http://www.tealeafnation.com/2013/03/chinese-users-resort-to-dark-humor-to-mask-fears-about-pig-carcases-in-shanghai-river

Millward, Steven, "8 Facts About Sina Weibo Users That All Marketers Should Know," *Tech in Asia*, November 16, 2012. As of November 26, 2014:
http://www.techinasia.com/sina-weibo-users-facts-marketers

Minter, Adam, "If You Thought China's Air Was Bad, Try the Water," *Bloomberg View*, April 14, 2014. As of January 14, 2015:
http://www.bloombergview.com/articles/2014-04-14/if-you-thought-china-s-air-was-bad-try-the-water

Moore, Malcolm, "China Kills off Discussion on Weibo After Internet Crackdown," *Telegraph*, January 30, 2014. As of November 26, 2014:
http://www.telegraph.co.uk/news/worldnews/asia/china/10608245/China-kills-off-discussion-on-Weibo-after-internet-crackdown.html

Nathan, Andrew J., "Authoritarian Resilience," *Journal of Democracy*, Vol. 14, No. 1, January 2003, pp. 6–17.

Ng, Jason Q., "Where Do Weibo Users Live? City and Provincial Breakdown of Various Chinese Internet Statistics," March 1, 2013. As of November 26, 2014:
http://blockedonweibo.tumblr.com/post/44289028375/where-do-weibo-users-live-city-and-provincial

———, "There Are Not Millions of Twitter Users in China: Supporting @ooof's Result and Refuting GWI's Conclusion," January 6, 2014. As of November 26, 2014:
http://blockedonweibo.tumblr.com/post/39828699303/there-are-not-millions-of-twitter-users-in-china

O'Brien, Kevin, and Lianjiang Li, *Rightful Resistance in Rural China*, Cambridge, UK: Cambridge University Press, 2006.

Olesen, Alexa, "China Eases Censorship On Food Safety Muckraking," Associated Press, May 16, 2011. As of May 19, 2015:
http://www.huffingtonpost.com/2011/05/16/china-press-freedom-food-safety_n_862448.html

Ong, Josh, "Report: Twitter's Most Active Country is China (Where It Is Blocked)" *The Next Web*, September 26, 2012. As of November 26, 2014:
http://thenextweb.com/asia/2012/09/26/surprise-twitters-active-country-china-where-blocked

Pennebaker, James W., and Cindy K. Chung, "Computerized Text Analysis of al-Qaeda Statements," in Klaus Krippendorff and Mary Angela Bock, eds., *A Content Analysis Reader*, Thousand Oaks, Calif.: Sage, 2008, pp. 453–466.

Pennebaker, J. W., C. K. Chung, M. Ireland, A. Gonzales, and R. J. Booth, "The Development and Psychometric Properties of LIWC2007," Austin, Texas: LIWC, Inc., 2007.

Pew Research Center, "Environmental Concerns on the Rise in China: Many Also Worried About Inflation, Inequality, Corruption," September 19, 2013. As of April 17, 2015:
http://www.pewglobal.org/2013/09/19/environmental-concerns-on-the-rise-in-china

Reuters, "China Regulator Says Scandal-Hit Food Supplier Forged Production Dates," July 26, 2014. As of November 26, 2014:
http://www.reuters.com/article/2014/07/26/china-food-investigation-idUSL4N0Q103T20140726

Roberts, Hal, Ethan Zuckerman, Jillian York, Robert Faris, and John Palfrey, "International Bloggers and Internet Control," Berkman Center for Internet and Society, August 2011.

Robinson, David, Harlan Yu, and Anne An, "Collateral Freedom: A Snapshot of Chinese Internet Users Circumventing Censorship," New York: Open Internet Tools Project, May 21, 2013.

Russell, Jon, "No, Facebook Does Not Have 63.5 Million Active Users in China," *The Next Web*, September 28, 2012. As of May 19, 2015:
http://thenextweb.com/asia/2012/09/28/no-way-jose

Schlaeger, Jesper, and Min Jiang, "Official Microblogging and Social Management by Local Governments in China," *China Information*, Vol. 28, No. 2, 2014, pp. 189–213.

Schmitz, Rob, "China's Fight for Cleaner Air," *Marketplace*, July 15, 2014. As of January 14, 2015:
http://www.marketplace.org/topics/sustainability/we-used-be-china/chinas-fight-cleaner-air

Shirk, Susan, "Changing Media, Changing China: Introduction," in Susan L. Shirk, ed. *Changing Media, Changing China*, Oxford, UK: Oxford University Press, 2011, pp. 1–37.

Sina Weibo, "Weibo Reports Second Quarter 2014 Financial Results," press release, August 14, 2014. As of November 26, 2014:
http://ir.weibo.com/phoenix.zhtml?c=253076&p=irol-newsArticle&ID=1958713&highlight

Sina Weibo Data Center, "2012 Sina Weibo Users Development Survey Report" ["2012 nian xinlang weibo yonghu fazhan diaocha baogao"], October 2012. As of November 26, 2014:
http://vdisk.weibo.com/s/AEdhAAT9K9k7

Sonnad, Nikhil, "Hacked Emails Reveal China's Elaborate and Absurd Internet Propaganda Machine," *Quartz*, December 18, 2014. As of January 14, 2015:
http://qz.com/311832/hacked-emails-reveal-chinas-elaborate-and-absurd-internet-propaganda-machine

"Stanford Word Segmenter," Version 3.4.1, August 27, 2014. As of May 19, 2015:
http://nlp.stanford.edu/software/segmenter.shtml

Sullivan, Jonathan, "A Tale of Two Microblogs in China," *Media, Culture and Society*, Vol. 34, No. 6, September 2012, pp. 773–783.

Tkacheva, Olesya, Lowell H. Schwartz, Martin C. Libicki, Julie E. Taylor, Jeffrey Martini, and Caroline Baxter, *Internet Freedom and Political Space*, Santa Monica, Calif.: RAND Corporation, RR-295-DOS, 2013. As of January 14, 2016:
http://www.rand.org/pubs/research_reports/RR295.html

Wong, Kam-Fai, Wenjie Li, Ruifeng Xu, and Zheng-sheng Zhang, *Introduction to Chinese Natural Language Processing*, San Rafael, Calif.: Morgan and Claypool Publishers, 2010.

World Bank, "GDP per capita, PPP (Current International $)," undated database. As of May 19, 2015:
http://data.worldbank.org/indicator/NY.GDP.PCAP.PP.CD

Xiao, Qiang, "The Rise of Online Public Opinion and Its Political Impact," in Susan L. Shirk, ed., *Changing Media, Changing China*, Oxford, UK: Oxford University Press, 2011, pp. 202–224.

Xu, Beina, "China's Environmental Crisis," Council on Foreign Relations Backgrounder, April 25, 2014. As of January 14, 2015:
http://www.cfr.org/china/chinas-environmental-crisis/p12608

Xue, Nianwen, Xiuhong Zhang, Zixin Jiang, Martha Palmer, Fei Xia, Fu-Dong Chiou, and Meiyu Chang, Chinese Treebank 8.0 LDC2013T21, web download, Philadelphia: Linguistic Data Consortium, 2013. As of May 19, 2015:
https://catalog.ldc.upenn.edu/LDC2013T21

Yeung, Douglas, and Astrid Stuth Cevallos, "The Mountains Are High and the Emperor Is Far Away," *Foreign Policy*, November 11, 2014. As of May 19, 2015: http://foreignpolicy.com/2014/11/11/the-mountains-are-high-and-the-emperor-is-far-away

Zhai, Ivan, "China May Only Have 18,000 Active Twitter Users: Infographic," *South China Morning Post*, January 3, 2013. As of November 26, 2014: http://www.scmp.com/comment/blogs/article/1119055/china-may-only-have-18000-active-twitter-users-infographic